219
Advances in Polymer Science

Advances in Polymer Science
Recently Published and Forthcoming Volumes

Self-Assembled Nanomaterials II
Nanotubes
Volume Editor: Shimizu, T.
Vol. 220, 2008

Self-Assembled Nanomaterials I
Nanofibers
Volume Editor: Shimizu, T.
Vol. 219, 2008

Interfacial Processes and Molecular Aggregation of Surfactants
Volume Editor: Narayanan, R.
Vol. 218, 2008

New Frontiers in Polymer Synthesis
Volume Editor: Kobayashi, S.
Vol. 217, 2008

Polymers for Fuel Cells II
Volume Editor: Scherer, G. G.
Vol. 216, 2008

Polymers for Fuel Cells I
Volume Editor: Scherer, G. G.
Vol. 215, 2008

Photoresponsive Polymers II
Volume Editors: Marder, S. R., Lee, K.-S.
Vol. 214, 2008

Photoresponsive Polymers I
Volume Editors: Marder, S. R., Lee, K.-S.
Vol. 213, 2008

Polyfluorenes
Volume Editors: Scherf, U., Neher, D.
Vol. 212, 2008

Chromatography for Sustainable Polymeric Materials
Renewable, Degradable and Recyclable
Volume Editors: Albertsson, A.-C., Hakkarainen, M.
Vol. 211, 2008

Wax Crystal Control · Nanocomposites Stimuli-Responsive Polymers
Vol. 210, 2008

Functional Materials and Biomaterials
Vol. 209, 2007

Phase-Separated Interpenetrating Polymer Networks
Authors: Lipatov, Y. S., Alekseeva, T.
Vol. 208, 2007

Hydrogen Bonded Polymers
Volume Editor: Binder, W.
Vol. 207, 2007

Oligomers · Polymer Composites Molecular Imprinting
Vol. 206, 2007

Polysaccharides II
Volume Editor: Klemm, D.
Vol. 205, 2006

Neodymium Based Ziegler Catalysts – Fundamental Chemistry
Volume Editor: Nuyken, O.
Vol. 204, 2006

Polymers for Regenerative Medicine
Volume Editor: Werner, C.
Vol. 203, 2006

Self-Assembled Nanomaterials I

Nanofibers

Volume Editor: Toshimi Shimizu

With contributions by

B.-K. Cho · Y.-W. Chung · A. Greiner · N. Higashi · H.-J. Kim
N. Kimizuka · T. Koga · B.-I. Lee · M. Lee · J. H. Wendorff

Springer

The series *Advances in Polymer Science* presents critical reviews of the present and future trends in polymer and biopolymer science including chemistry, physical chemistry, physics and material science. It is adressed to all scientists at universities and in industry who wish to keep abreast of advances in the topics covered.
As a rule, contributions are specially commissioned. The editors and publishers will, however, always be pleased to receive suggestions and supplementary information. Papers are accepted for *Advances in Polymer Science* in English.
In references *Advances in Polymer Science* is abbreviated *Adv Polym Sci* and is cited as a journal.

Springer WWW home page: springer.com
Visit the APS content at springerlink.com

ISBN 978-3-642-09878-9 e-ISBN 978-3-540-85103-5
DOI 10.1007/978-3-540-85103-5

Advances in Polymer Science ISSN 0065-3195

Softcover reprint of the hardcover 1st edition 2008

Cover design: WMXDesign GmbH, Heidelberg

Printed on acid-free paper

9 8 7 6 5 4 3 2 1 0

springer.com

Volume Editor

Prof. Dr. Toshimi Shimizu
Nanoarchitectonics Research Center (NARC)
National Inst. of Advanced Industrial Science and Technology (AIST)
1-1-1 Higashi, Tsukuba
Ibaraki 305-8565, Japan
tshmz-shimizu@aist.go.jp

Editorial Board

Advances in Polymer Science
Also Available Electronically

For all customers who have a standing order to Advances in Polymer Science, we offer the electronic version via SpringerLink free of charge. Please contact your librarian who can receive a password or free access to the full articles by registering at:

springerlink.com

If you do not have a subscription, you can still view the tables of contents of the volumes and the abstract of each article by going to the SpringerLink Homepage, clicking on "Browse by Online Libraries", then "Chemical Sciences", and finally choose Advances in Polymer Science.

You will find information about the

- Editorial Board
- Aims and Scope
- Instructions for Authors
- Sample Contribution

at springer.com using the search function.

Color figures are published in full color within the electronic version on SpringerLink.

Preface

Nanotechnology is the creation of useful materials, devices, and systems through the control of matter on the nanometer-length scale. This takes place at the scale of atoms, molecules, and supramolecular structures. In the world of chemistry, the rational design of molecular structures and optimized control of self-assembly conditions have enabled us to control the resultant self-assembled morphologies having 1 to 100-nm dimensions with single-nanometer precision. This current research trend applying the bottom-up approach to molecules remarkably contrasts with the top-down approach in nanotechnology, in which electronic devices are miniaturizing to smaller than 30 nm. However, even engineers working with state-of-the-art computer technology state that maintaining the rate of improvement based on Moore's law will be the most difficult challenge in the next decade.

On the other hand, the excellent properties and intelligent functions of a variety of natural materials have inspired polymer and organic chemists to tailor their synthetic organic alternatives by extracting the essential structural elements. In particular, one-dimensional structures in nature with sophisticated hierarchy, such as myelinated axons in neurons, tendon, protein tubes of tubulin, and spider webs, provide intriguing examples of integrated functions and properties.

Against this background, supramolecular self-assembly of one-dimensional architectures like fibers and tubes from amphiphilic molecules, bio-related molecules, and properly designed self-assembling polymer molecules has attracted rapidly growing interest. The intrinsic properties of organic molecules such as the diversity of structures, facile implementation of functionality, and the aggregation property, provide infinite possibilities for the development of new and interesting advanced materials in the near future. The morphologically variable characteristics of supramolecular assemblies can also function as pre-organized templates to synthesize one-dimensional hybrid nanocomposites. The obtained one-dimensional organic–inorganic, organic–bio, or organic–metal hybrid materials are potentially applicable to sensor/actuator arrays, nanowires, and opto-electric devices.

The present volumes on Self-Assembled Nanofibers (Volume 219) and Nanotubes (Volume 220) provide an overview on those aspects within eight chapters. Different points of view are reflected, featuring interesting aspects related to (a) the self-assembly of supramolecular nanofibers comprising of organic, polymeric, inorganic and biomolecules (N. Kimizuka, in Volume 219, Chapter 1), (b) controlled self-assembly of artificial peptides and peptidomimetics into nanofiber architectures (N. Higashi, T. Koga, in Volume 219, Chapter 2), (c) self-assembled nanostructures from amphiphilic rod molecules (B.-K. Cho, H.-J. Kim, Y.-W. Chung, B.-I. Lee, M. Lee, in Volume 219, Chapter 3), (d) the production of functional self-assembled nanofibers by electrospinning (A. Greiner, J. H. Wendorff, in Volume 219, Chapter 4), (e) the synthesis of tailored π-electronic organic nanotubes and nanocoils (T. Yamamoto, T. Fukushima, T. Aida, in Volume 220, Chapter 1), (f) preparation and fundamental aspects of nanotubes self-assembled from block copolymers (G. Liu, in Volume 220, Chapter 2), (g) β-1,3-glucan that can act as unique natural nanotubes and incorporate conjugated polymers or molecular assemblies (M. Numata, S. Shinkai, in Volume 220, Chapter 3), and (h) the fabrication of self-assembled polymer nanotubes involving the use of a nanoporous hard template (M. Steinhart, in Volume 220, Chapter 4). A variety of nanofibers and nanotubes with well-defined morphologies and dimensions are discussed in terms of self-assembly of molecular and polymer building blocks in bulk solution or confined geometry like nanopores.

Current materials and manufacturing technologies strongly require technological advances for reducing environmental load combined with energy and resource savings in production. In order to develop such technologies for the development of a sustainable society, research on materials production based on the self-assembly technique is of great interest. Hopefully, these volumes will be beneficial to readers involved with self-organization in the field of bottom-up nanotechnology as well as those concerned with industrial fiber processing.

Tsukuba, June 2008 Toshimi Shimizu

Contents

Self-Assembly of Supramolecular Nanofibers
N. Kimizuka 1

Self-Assembled Peptide Nanofibers
N. Higashi · T. Koga 27

Self-Assembled Nanofibers
and Related Nanostructures from Molecular Rods
B.-K. Cho · H.-J. Kim · Y.-W. Chung · B.-I. Lee · M. Lee 69

Functional Self-Assembled Nanofibers by Electrospinning
A. Greiner · J. H. Wendorff 107

Subject Index 173

Contents of Volume 220

Self-Assembled Nanomaterials II

Nanotubes

Volume Editor: Shimizu, T.
ISBN: 978-3-540-85104-2

Self-Assembled Nanotubes and Nanocoils from π-Conjugated Building Blocks
T. Yamamoto · T. Fukushima · T. Aida

Block Copolymer Nanotubes Derived from Self-Assembly
G. Liu

Self-Assembled Polysaccharide Nanotubes Generated from β-1,3-Glucan Polysaccharides
M. Numata · S. Shinkai

Supramolecular Organization of Polymeric Materials in Nanoporous Hard Templates
M. Steinhart

Adv Polym Sci (2008) 219: 1–26
DOI 10.1007/12_2008_170

Published online: 15 August 2008

Self-Assembly of Supramolecular Nanofibers

Nobuo Kimizuka

Department of Chemistry and Biochemistry, Graduate School of Engineering, Kyushu University, JST CREST, 744 Moto-oka, Nishi-ku, 819-0395 Fukuoka, Japan
n-kimi@mail.cstm.kyushu-u.ac.jp

1 **Introduction** 2

2 **Nanofibers Self-Assembled from Classical Amphiphiles** 3

3 **Self-Assembly of Supramolecular Nanofibers in Aqueous Media** 5
3.1 Molecular Recognition-Based Approach 5
3.2 Nanofibers Formed by Molecular Pairing 7
3.3 Nanofibers Formed by Biological Molecular Components 9

4 **Organic/Inorganic Nanofibers Self-Assembled in Organic Media** 12
4.1 Pseudo One-Dimensional Halogen-Bridged Mixed Valence Complexes . . . 12
4.2 Formation of 1D Conjugated Electronic Structures in Response to Self-Assembly 14
4.3 Supramolecular Band Gap Engineering and Solvatochromic Nanowires . . 15
4.4 Self-Assembly of Nanofibers at Interfaces 17
4.5 Nanofibers Self-Assembled from Lipophilic Bridging Ligands and Metal Ions 18
4.6 Supramolecular Control of Nanofiber Morphology and Spin Crossover . . 20

5 **Conclusion and Outlook** 23

References 23

Abstract The development of functional nanofibers via self-assembly is an area of growing interest. Whereas self-assembly has been successful in creating nanostructures, the fabrication of hierarchical architectures from multiple functional components and the development of synergistic functions are emerging as challenging issues. In this chapter, various self-assembly approaches that are used to construct nanofibers are summarized. Self-assembly of amphiphilic supermolecules provides a key strategy to hierarchically fabricate multicomponent architectures. Amphiphilic pairs of nanofiber-forming molecules can be screened from a wide range of functional molecules, including biomolecules. As examples of nanofibers whose functions are dynamically controlled based on self-assembly, lipophilic one-dimensional metal complexes and their unique characteristics are introduced.

Keywords Charge Transfer · Hydrogen-bonding · Mesoscopic supramolecular assembly · Metal complex · Nanofiber · Self-assembly · Spin crossover

1 Introduction

One dimensional nanostructures have been gathering much attentions as building blocks for nanodevices. Biological systems are replete with such nanoarchitectures which are spontaneously formed by ingeniously employing noncovalent interactions [1]. One of the most elaborated supramolecular fibers with pivotal biological significance is neurons, whose fundamental task is to receive, conduct and transmit signals. Neurons are typically composed of a cell body called soma, a dendritic tree and an axon (Fig. 1) [2]. The axons of many vertebrate neurons are wrapped by the plasma membrane of glial cells, which insulate the axonal membranes and prevent the current leaks across them. This supramolecular packaging enables action potentials to travel faster than in unmyelinated axons.

The fine superarchitecture of myelinated axons inspires us to design a new generation of supramolecular fibers. First, they clearly indicate the importance of hierarchical architectures, where heterogeneous molecular components or supramolecular assemblies are self-organized to perform integrated functions. In contrast to conventional nanofiber structures formed from identical molecular components, molecular organization of such heterogeneous molecular components would allow the design of synergistic functions. The components of supramolecular nanofibers can be selected widely from organic, polymeric, inorganic and bio-molecules (Fig. 2). Each categories – organic, inorganic and biomolecules – are replete with materials classified according to chemical structures and properties, and molecular components can be selected from each category. There are uninvestigated combinations of molecular components especially in the interdisciplinary area (I–IV), in which a wealth of new phenomena and functions are expected. Ordered

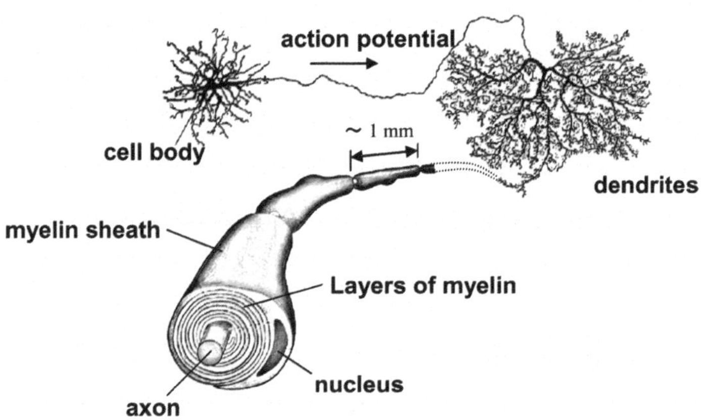

Fig. 1 Schematic illustration of a myelinated axon from a peripheral nerve [2]

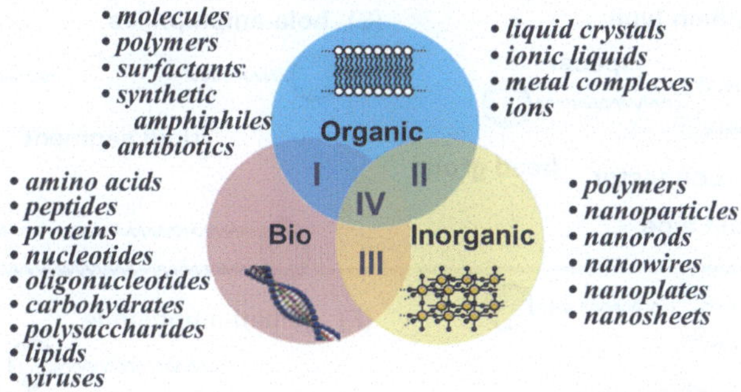

Fig. 2 Three categories of supramolecular components and interdisciplinary area I–IV

self-assembly of hetero-components inevitably produces molecular or nano-interfaces; this would exert decisive roles in determining the property and function of molecular systems. Second, imparting responsive and adaptive natures are important to develop smart molecular systems whose functions are expressed in harmony with the environmental or physicochemical stimuli. They would provide a basis to design neuromorphic properties, as exemplified by self-wiring and self-restoration. They may also lead to molecular systems with self-learning functionalities [3]. In this chapter, progress in the self-assembly of supramolecular nanofibers will be reviewed especially from these perspectives.

2 Nanofibers Self-Assembled from Classical Amphiphiles

Nanofibers are self-assembled from properly designed synthetic amphiphiles in aqueous or in organic environments [4]. In addition to nanofibers, a wide variety of mesoscopic-scale structures such as vesicles, tubes, disks, lamellas, and helices are formed from suitably designed amphiphiles – bolaamphiphiles, gemini surfactants in addition to the classical amphiphiles. These amphiphilic systems contain a polar head group and one, two or multiple hydrophobic tails (Fig. 3) [4–10]. The aqueous systems that assemble into discrete, nanofibers from dilute solution usually require sufficient amphiphilicity, and such condition is nowadays expanded to a variety of synthetic amphiphiles as exemplified by peptide-amphiphiles [11–14], oligopeptides [15–17], block copolymers [18–20] and rod-like molecules [21–23]. These nanofibers show extremely diverse properties and were shown to exert potentials in numerous bio-nanotechnologies. Increase in the inter-nanofiber interactions causes thickening of dispersions and gelation, where amphiphilic molecular designs for hydrogelators [24] and organogelators [25, 26] are more tolerated compared to the dispersed systems.

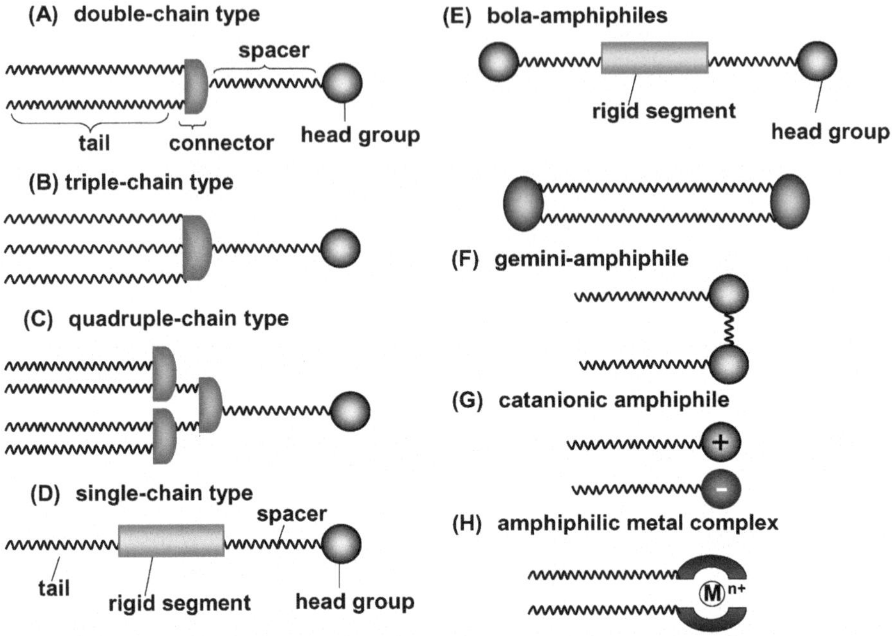

Fig. 3 Schematic chemical structures of bilayer-forming amphiphiles

The soluble nanofiber-forming systems discussed here are single amphiphilic molecules that are comprised of a solvophilic group and solvophobic group connected via covalent bonding. However, they can be also designed from heterogeneous molecular components, which interact with each other through non-covalent interactions. To design soluble nanofibers, interaction of nanofiber surfaces with solvent molecules needs to be considered. In this context, the popular strategy of supramolecular polymers, defined as: "polymeric arrays of repeating molecular units that are assembled by reversible and directional non-covalent interactions" [27] does not give sufficient guidelines, since it only refers to their primary structures. Similar to the protein folding in aqueous environment, interactions with the environment determine their secondary and ternary structures. As described in the introduction, self-organization includes the interaction with environments. Therefore, a comprehensive concept of mesoscopic supramolecular assemblies would be defined as "*hierarchically self-assembled amphiphilic supermolecules whose ternary and higher assembly structures are controlled through the solvophilic–solvophobic interactions*" [10]. Here, pairs of molecules brought by secondary interactions are designed, which acquire amphiphilicity upon complexation. They become units of self-assembly, and hierarchically grow into mesoscopic-scale supermolecules which are dispersed stably in aqueous or organic media.

3
Self-Assembly of Supramolecular Nanofibers in Aqueous Media

3.1
Molecular Recognition-Based Approach

Amphiphilic structures can be formed from combination of molecular components, which interact with each other via non-covalent interactions. A classical example of such hybrid amphiphiles is "catanionic surfactants" (Fig. 3G) [28, 29]. Mixtures of anionic and cationic surfactants display varied aggregate structures, such as globular and rodlike micelles, vesicles and lamellar phases. Amphiphilic metal complexes "metalloamphipihles" are formed from metal ions and lipophilic ligand molecules (Fig. 3H) [30, 31]. Thus, electrostatic interactions and coordination bonding have been employed to design hybrid amphiphiles. In contrast to supramolecular complexes formed via complementary hydrogen bonding in aprotic organic media [32–35], hydrogen bonding has not been employed as assembly directing interactions in water. This is because enthalpic gain by the formation of hydrogen bonds in water is cancelled by the enthalpy to break hydrogen bonds, which priorly existed between these molecules and water [36]. To compensate the entropic disadvantage, hydrogen bonding in the aqueous media requires integration of the other noncovalent interactions – such as hydrophobic interactions or aromatic stacking found in DNA double

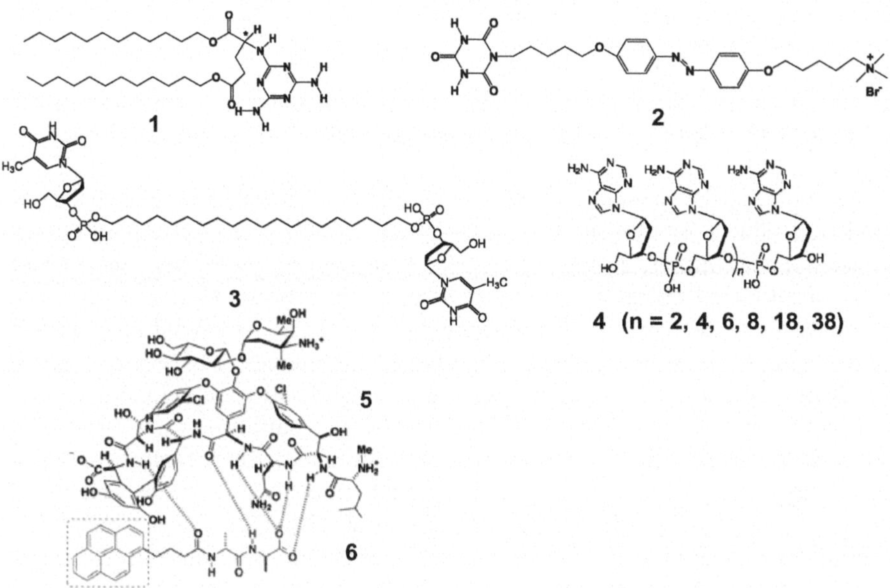

Fig. 4 Specific molecular pairs that give amphiphilic supermolecules

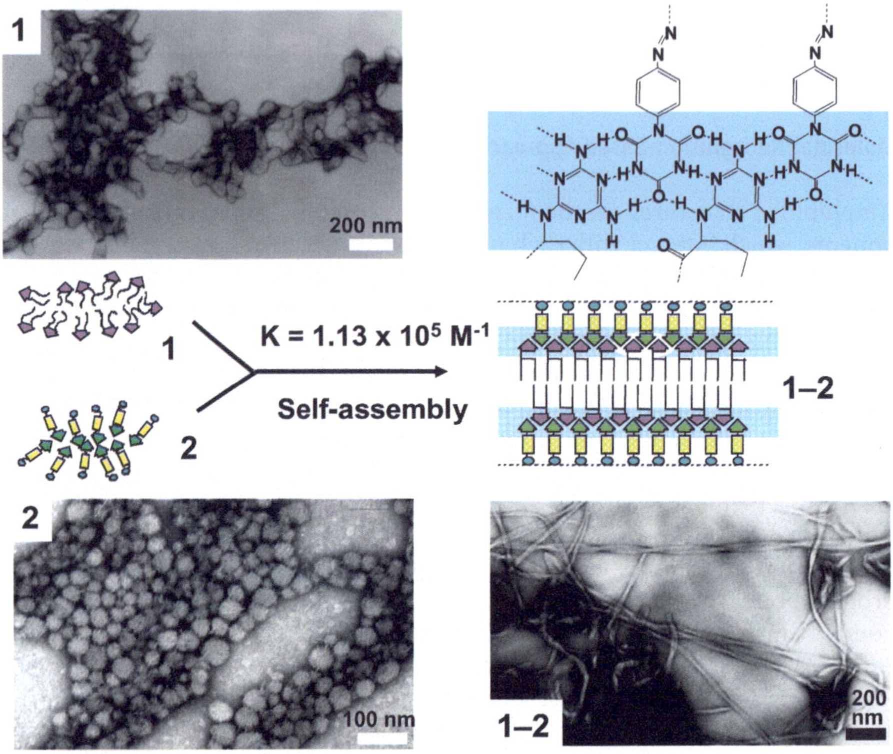

Fig. 5 Spontaneous self-assembly of **1** and **2** in water gives helical nanofibers consisting of supramolecular membranes **1–2** [37]

helices. In these duplexes, complementary base pairs are formed inside, forming a hydrophobic column of base pairs that are stabilized by stacking [2].

By applying the principle of nucleic acid hybridization, a hydrogen-bond-mediated bilayer membrane can be designed from a pair of complementary molecular subunits **1**, **2** (Figs. 4 and 5). When amphiphilic hydrogen bond pairs of alkylated chiral melamine (hydrophobic subunit **1**) and quaternary ammonium - derivatized azobenzene cyanuric acid (hydrophilic subunit **2**) were mixed in water, helical nanofibers (thicknesses: 14–28 nm, widths: 30–50 nm, pitches: 180–430 nm) were immediately formed (Fig. 5, association constant: $1.13 \times 10^5\,M^{-1}$) [37]. These helical nanofibers are typical of the ordered chiral bilayer membranes [38] and are distinct from the aggregate morphology observed for the individual subunits. It is noteworthy that the presence of water directs the self-assembly. The hydrophobic melamine subunits are located in the interior, and the ammonium-containing counterparts constitute the hydrophilic surface of the assemblies. Among the possible combinations of complementary hydrogen bond networks, such as circular (rosette), crinkled tape, and linear tape [32], only the linear network structure was selected since it provides the most amphiphilic superstructure. This feature is common to the supramolecular hybridization of biological polymers and also to the protein folding. It provided the first example of artificial nanofibers self-assembled from two complementary subunits via hydrogen bonding in water.

Shimizu et al. reported formation of helical nanofibers with a width of 7–8 nm by mixing nucleotide-appended bolaamphiphile **3** and oligoadenylic acids **4** [39]. The nanofiber length and their thermal stability are increased when longer oligoadenylic acids are employed, suggesting the assembly is templated by the oligonucleotide. Molecular recognition – driven self-assembly of bio-active small molecules has also been demonstrated by Xu et al. [40]. Vancomycin **5**, clinically important antibiotics in treating Gram-positive bacterial infections, was employed as a receptor, and mixing of a pyrene-conjugated D-Ala-D-Ala derivative **6** in water gave hydrogel at the minimum gel concentration of 5 mM. This concentration is lower than the nanofibrous hydrogel formed by **6** alone (ca. 30 mM) and storage modulus of the gel was raised by 10^6 fold for the binary mixture **5** + **6**. Electron microscopy of the **5** + **6** hydrogel give sheetlike two-dimensional structures, and host–guest interactions caused extensive cross-linking of the complexes.

3.2 Nanofibers Formed by Molecular Pairing

Nanofibers are also formed in water by amphiphilic ion pairs. Huc et al. showed formation of helical nanofibers and hydrogels from Gemini surfactants **7** with L-tartrate and D-tartrate counterions (Fig. 6) [41]. These chiral structures are originating from tartrate molecules, and their molecular orientation and hydrogen bonding at the membrane surface are responsible the morphology. The ion pairs of dialkyldimethylammonium (**8**, $n = 12, 14$) and guanosine 5′-monophosphate (GMP), adenosine 5′-monophosphate (AMP) were recently shown to form nano to micrometric left-handed helices in water [42].

The electrostatic molecular pairing approach can be used to develop functional nanofibers. Inspired by the peptide-assisted organization of chlorophylls in photosynthetic systems, light-harvesting supramolecular hydrogels were developed [43]. Cationic L-glutamate derivatives **9**, **10** are designed as self-assembling scaffolds which bind anionic fluorophores through the electrostatic and van der Waals interactions. Compounds **9** and **10** possess shorter alkyl chains compared to the conventional bilayer-forming amphiphile **11**. Such short alkyl chains prevent precipitation upon the formation of hydrophobic ion pairs. Equimolar mixtures of **9**/NaphSO$_3^-$ (3 mM), **10**/NaphSO$_3^-$ and **10**/AnthSO$_3^-$ (10 mM) spontaneously formed hydrogels as

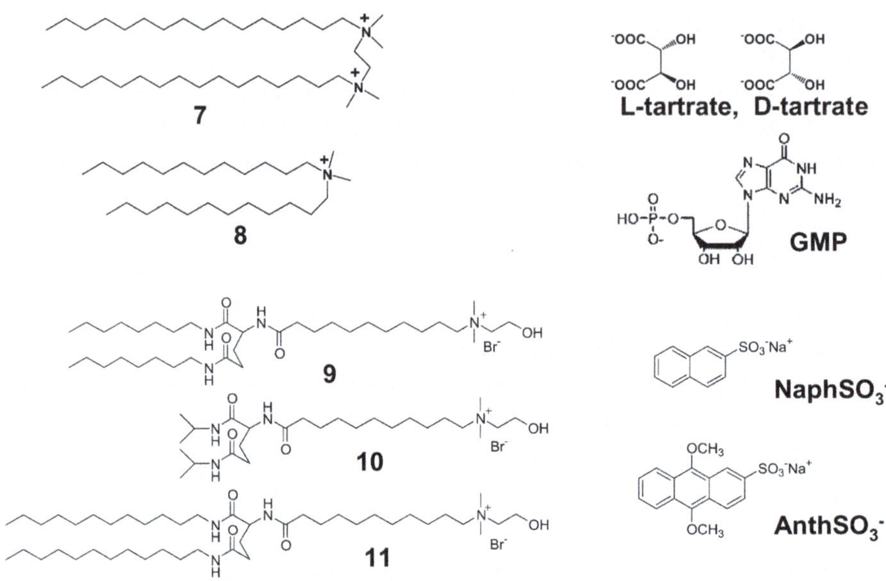

Fig. 6 Chemical structures of nanofiber-forming ion pairs

schematically shown in Fig. 7. They gave bundled nanofibers which are physically crosslinked, due to reduced electrostatic repulsions between the fibers. Interestingly, hydrogel of **9**/$NaphSO_3^-$ showed efficient energy transfer characteristics.

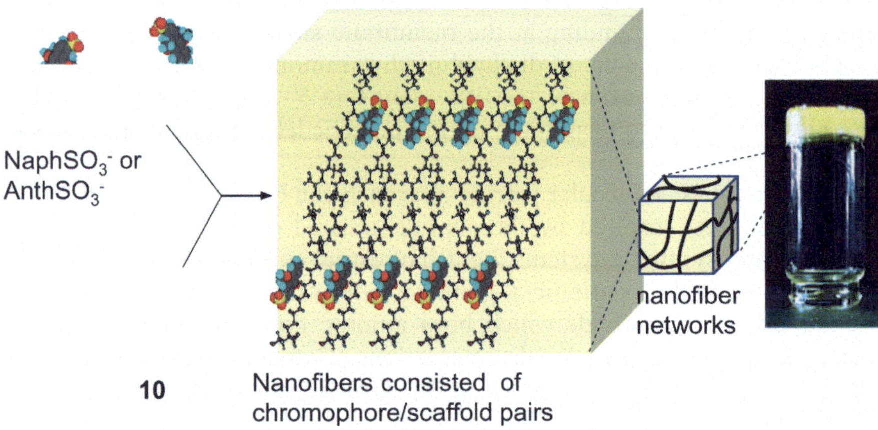

Fig. 7 Light-harvesting hydrogels self-assembled from anionic fluorophores and cationic molecular scaffolds [43]

3.3 Nanofibers Formed by Biological Molecular Components

Hetero-component nanofibers self-assembled in aqueous media have been more recently extended to biologically important molecules. To date, studies on the self-assembly of biological molecules have been focused on biopolymers such as proteins [44], oligopeptides [16] and nucleic acids [45]. Smaller biomolecules have been important targets for bio- and chemosensors, but they have not been considered as building blocks for creating functional supramolecular systems. For example, ATP, a universal energy currency in all of the biological systems, has been recognized by synthetic host–guest receptors [46], peptides [47] and RNA aptamers [48], development of ATP-directed artificial self-assembly has been unexplored. We have searched for cationic chromophores that display color changes in the presence of ATP and came across a thiacarbocyanine dye **12**. When ATP was added to an aqueous solution of **12** at a varied molar ratio, immediate color change occurred from pink to orange [49]. Induced circular dichroism (ICD) with exciton coupling was observed for the absorption band of achiral chromophores, indicating excitonic interactions among them. In transmission electron microscopy, developed nanofibers with a minimum width of ca. 10 nm and lengths of several μm were observed (Fig. 8b). The observed width of nanofibers is larger than the molecular size of **12** (ca. 1.6×0.7 nm) and ATP (ca. 1.4×1.0 nm), indicating that they consist of bundled **12**/ATP complexes. The ATP-directed self-assembly is thermally reversible. The excitonic nanofibers formed at room temperature are dissociated into non-aggregated ion pairs at elevated temperatures, while upon cooling nanofibers with regular widths of ca. 40 nm are reassembled. The binding ratios of **12** : ATP as determined spectrophotometrically was 6 : 1, which is deviated from the value expected from electrostatic interactions. The excess binding of **12** to ATP indicates that multiple secondary forces such as van der Waals interactions, hydrophobic interactions and aromatic stacking are involved in the self-assembly, in addition to the electrostatic interactions.

The above molecular pairing approaches, however, were not applicable to small, neutral ionic molecules, such as zwitterionic amino acids. Though the molecular structure of amino acids (i.e., chirality, structure of side chains) is fundamental molecular information in the study of biology, molecular recognition of small amino acid residues remains difficult, and there exist no general strategy to amplify their structural information by using self-assembly. To extend the heterogeneous molecular pairing approach to amino acids, an in situ pre-modification technique has recently been developed (Fig. 8c) [50]. Amino acids are converted to isoindole derivatives (I-amino acid) by the reaction with ortho-phthalaldehyde (OPA) and alkylthiols. 2-Mercaptoethanesulfonic acid (MES) was employed since it gives an anionic charge tethered to the isoindole unit. Upon mixing amino acids, OPA

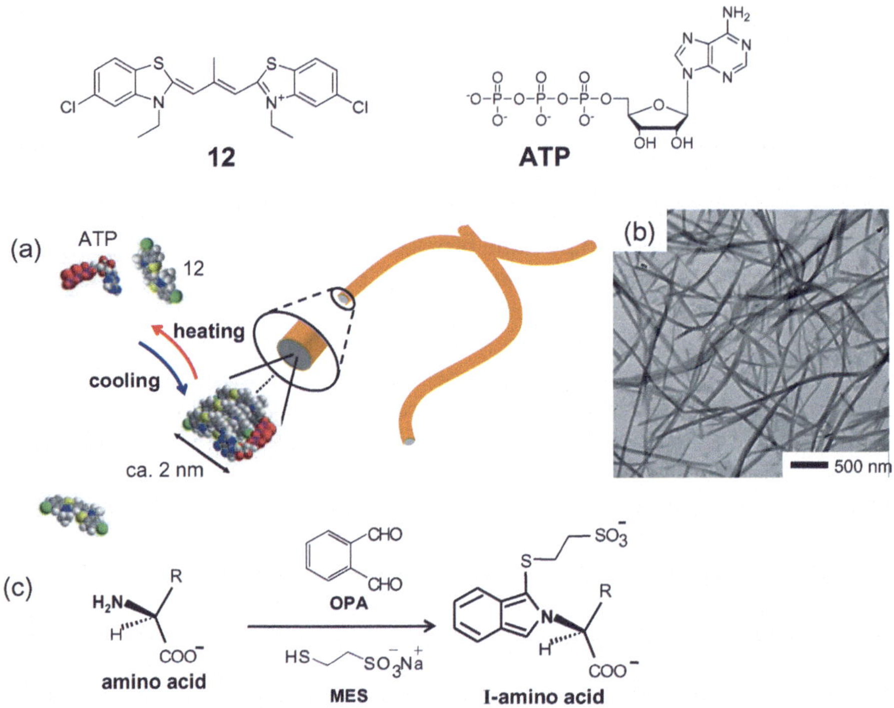

Fig. 8 **a** Schematic illustration of nanofibers self-assembled from ATP and cyanine dye **12** [49]. **b** Transmission electron microscopy of ATP/**12** nanofibers **c** one-pot conversion of aminoacids to isoindole-derivatives [50]

and MES in water, I-amino acids are immediately formed, as confirmed by the appearance of fluorescence and a new absorption peak at 333 nm. Introduction of isoindole and anionic groups enhanced interactions with cationic molecules, since addition of cyanine dye **12** to aqueous I-amino acids caused immediate color changes from pink to reddish pink or orange, depending on the amino acid employed. These color changes were not observed when amino acids and cyanine dyes **12** were mixed in the absence of OPA and MES. Therefore, the observed color changes originated from interactions between I-amino acids and dyes. In transmission electron microscopy (TEM), developed nanofibers are abundantly observed for I-Glu/**12** and I-Ala/**12** (width, ca. 50 nm. length, several μm). Interestingly, I-Glu/**12** and I-Ala/**12** showed intense induced circular dichroism (ICD) spectra with complex exciton coupling patterns. It is notable that the difference in chemical structure of side chains and chirality was readily distinguishable by ICD patterns. Moreover, the intensity of these visible ICD peaks is enhanced by more than 100 times compared to CD spectral intensity of I-amino acids in the ultraviolet region. Therefore, molecular information of amino acids is amplified and translated in UV-Vis absorption and circular dichroism, through self-assembly.

These observations indicate that biologically important small molecules turn to building blocks for the functional nanoarchitectures. It should be emphasized that the chemical structure of components and combination of multiple interactions, van der Waals forces, hydrophobic interactions, coordination interactions, hydrogen bonds in addition to the long-ranged electrostatic interactions, determine the composition and molecular orientation in the heteromolecular nanoassemblies.

Electrostatic complexation of polyelectrolytes, a series of charged dyes with oppositely charged surfactants has been investigated by Faul and Antonietti et al. [51–54]. As neutralization of electric charges and formation of polyioncomplexes tend to dehydrate the ion pairs, nondirectional secondary aggregation often lead to precipitation of the complexes in water. However, by employing charged tectons like cationic perylenediimide **13** and copper-phtalocyanine tetrasulfonate **14**, nanofibers with an average thickness of 2.4 nm were formed, as observed by cryo-transmission electron microscopy (Fig. 9) [54]. The composition of the complex was found to be $(\mathbf{13})_2(\mathbf{14})$, and helically arranged dye tectons in fiber-like structure was proposed (Fig. 9a). Shelnutt et al. reported formation of porphyrin nanofibers in dichloromethane from the combination of cationic oxo-antimony(V) por-

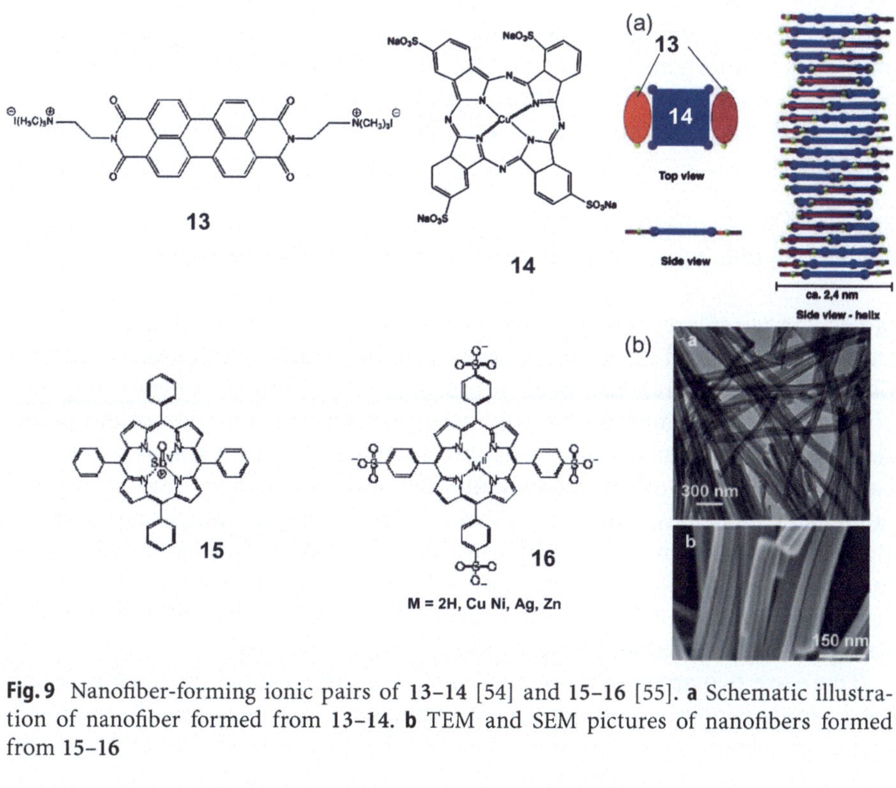

Fig. 9 Nanofiber-forming ionic pairs of **13–14** [54] and **15–16** [55]. **a** Schematic illustration of nanofiber formed from **13–14**. **b** TEM and SEM pictures of nanofibers formed from **15–16**

phyrin **15** and anionic porphyrins **16**. Bundles of nanofibers with diameter of 20–40 nm were observed by TEM and SEM (Fig. 9b) [55].

4 Organic/Inorganic Nanofibers Self-Assembled in Organic Media

Formation of supramolecular polymers in organic media have been extensively reviewed, especially for complementary hydrogen bonded systems [10, 34, 35, 56, 57]. To avoid inclusion of duplicate information, we focus here on the nanofibers self-assembled from organic/inorganic components in organic media. To date, a variety of supramolecular architectures have also been created by the use of metal coordination. Among them, synthesis of catenanes, knots [58] and double-helical metal complexes (helicates) [59] are the prominent examples. Nanostructured supramolecular squares and capsules are also formed by self-assembly [60]. These supramolecular complexes have been attracting widespread interest, however they consist of *discrete* metal complexes. Consequently, their applications are largely limited to the shape-related functions such as host–guest inclusion. It is only recently that interactions among the connecting metal ions at low temperatures are being investigated [61]. At the same time, the control on the magnetism and electronic states in the solid metal complexes represent other important issues in inorganic chemistry [62]. It would be natural to envisage that nanofibers in the next generation would integrate these two features, and conjugated electronic structures would be generated in response to the self-assembly.

4.1 Pseudo One-Dimensional Halogen-Bridged Mixed Valence Complexes

Molecular wires are indispensable elements of future molecular-scale electronic devises, and their fabrication has been attracting much interest. Conventional research has been focused on the synthesis of π-conjugated oligomers and supramolecular polymers [56]. The introduction of inorganic elements or one-dimensional inorganic complexes in main chains would expand the potential of these nanomaterials and lead to new interfacial phenomena, which have not been observed in the single component system. However, one-dimensional complexes have not been considered candidates for molecular wires, since they exist only in three-dimensional solids. They are usually not soluble, and when they are dissolved, the one-dimensional structure is disrupted and the structure is not maintained.

There are two approaches to solubilize one-dimensional metal complexes. We have developed a supramolecular strategy to package one-dimensional metal complexes by synthetic lipid molecules [63–74]. The other approach is to employ lipophilic bridging ligands [75–80]. In the first approach, we have focused on a family of quasi-one-dimensional, halogen-bridged M^{II}/M^{IV} mixed valence complexes $[M^{II}(en)_2][M^{IV}X_2(en)_2]Y_4$ (M = Pt, Pd, Ni, X = Cl, Br, I; Y : counterions, such as ClO_4^-) (Fig. 10). They have been attracting interests among solid-state physists and chemists, because of their strong electron-lattice interactions and unique physicochemical properties, such as intense intervalence charge-transfer (CT) absorption, semiconductivity, and large third-order nonlinear optical susceptibilities [81, 82]. In these compounds, the 1-D electronic structure is comprised mainly of d_z^2 orbital of the metals and p_z orbital of the bridging halogens, where z axis is parallel to the chain. The ground state can be expressed as a mono-valence state (spin density wave state, SDW) or a mixed-valence state (charge density wave state, CDW), where SDW state has only been reported for the Ni compounds [83]. In the case of $[Pt^{II}(en)_2][Pt^{IV}X_2(en)_2]$ compounds, the halogen bridging anions are located nearby $Pt^{IV}(en)_2$, with an off-center displacement of δ (Fig. 10c). This structural change is ascribed to the Peierls transition, which is a characteristic of the one-dimensional electronic systems.

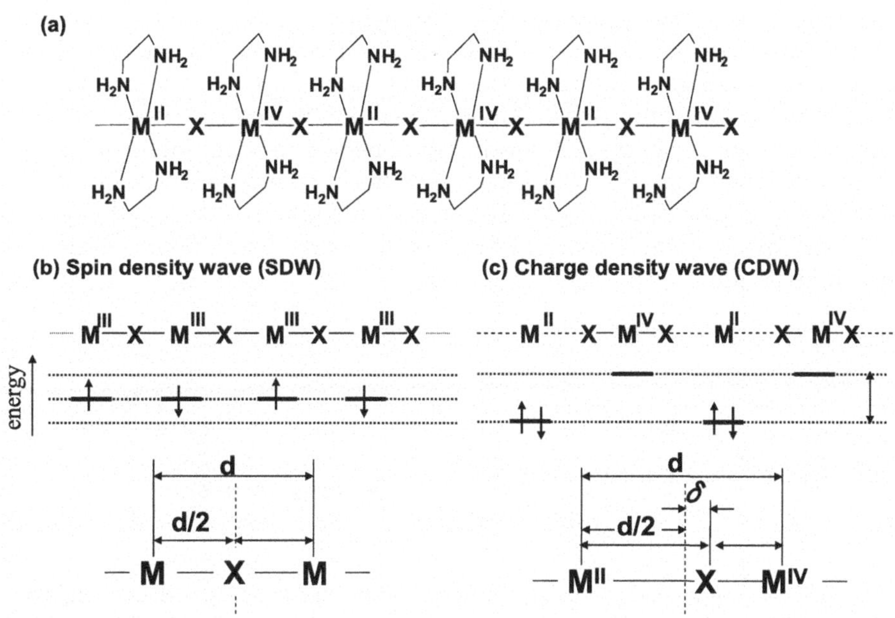

Fig. 10 **a** Schematic structure of halogen-bridged mixed valence complex $[M(en)_2][MX_2(en)_2]_4^{4+}$ (M = Pt, Pd, Ni; X = Cl, Br, I). Counteranions are omitted for clarity. **b** Spin density wave state (reported for M = Ni), **c** Charge density wave state (M = Pt, Pd etc.)

4.2
Formation of 1D Conjugated Electronic Structures in Response to Self-Assembly

To convert the solid complex $[Pt^{II}(en)_2][Pt^{IV}X_2(en)_2](ClO_4)_4$ to lipophilic structures, we have introduced lipid counteranions (Fig. 11) and prepared ternary inorganic/organic polyion complexes $[Pt(en)_2][PtCl_2(en)_2](lipid)_4$ as schematically shown in Fig. 12. $[Pt(en)_2][PtCl_2(en)_2]$(**17**, **18**, **19**, **20** or **21**)$_4$ displayed yellow (lipid = **17**, **20**) or indigo colors (lipid = **18**, **19**, **21**), depending on the chemical structure of amphiphiles [63–73]. These colors are typical of the intervalence charge-transfer (CT: $Pt^{II}/Pt^{IV} \rightarrow Pt^{III}/Pt^{III}$) of halogen-bridged linear complexes. The $[Pt(en)_2][PtCl_2(en)_2](lipid)_4$ complexes can be dispersed in organic media with the maintenance solid-state colors, indicating that the one-dimensional Pt^{II}–Pt^{IV} complexes are maintained in organic media with the help of solvophilic alkyl chains [65].

Interestingly, in the case of the supramolecular complex consisting of dihexadecyl sulfosuccinate **18** and $[Pt(en)_2][PtCl_2(en)_2]^{4+}$, indigo-colored dispersion was obtained at room temperature, whereas the color disappeared upon heating the solution to 60 °C. As the CT transition requires the existence of a chloro-bridged extended coordination structure, the observed thermochromism indicates disruption of the one-dimensional complex structure at 60 °C. The color reappeared reversibly upon cooling and the observed thermochromism indicates that the one-dimensional complex undergoes reversible re-assembly in solution (Fig. 12, supramolecular thermochromism). A TEM picture of $[Pt(en)_2][PtCl_2(en)_2]$(**18**)$_4$ in chloroform, after the heat treatment is shown in Fig. 12, where nanofibers with a minimum width of 18 nm and lengths of 700–1700 nm are seen in abundance [65]. The observed widths of the nanostructures are larger than the bimolecular lengths of **18**, and therefore they must consist of aggregates of the amphiphilic supramolecular structure.

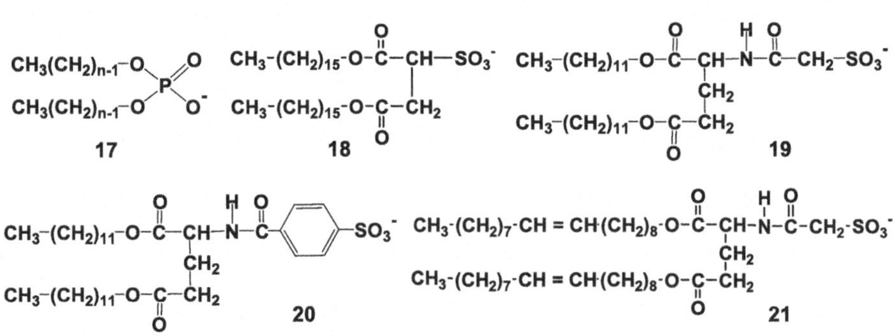

Fig. 11 Chemical structure of anionic amphiphiles used to package metal complexes

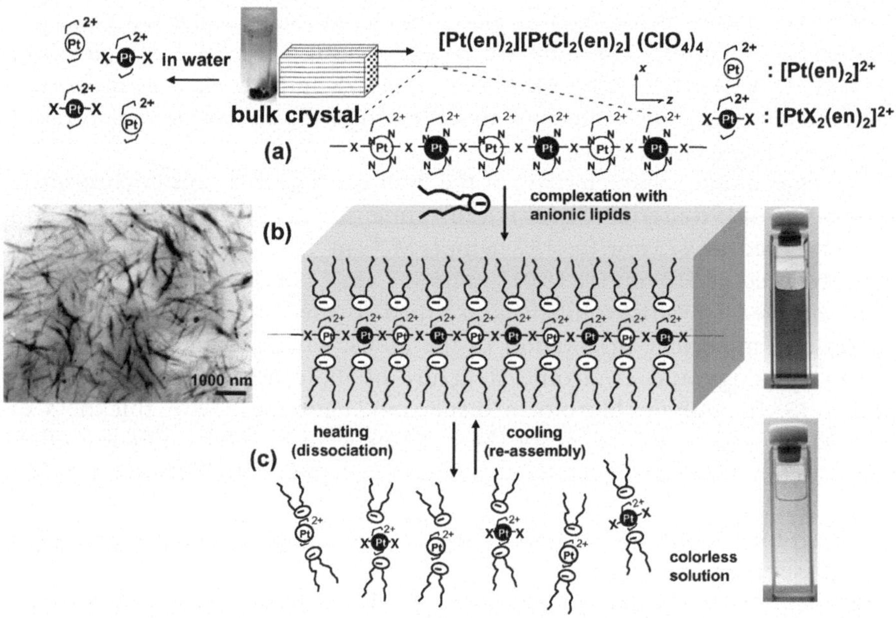

Fig. 12 Schematic illustration of self-assembling nanowires and supramolecular thermochromism [65]. **a** Schematic illustration for halogen-brideged mixed valence complex $[Pt(en)_2][PtX_2(en)_2]$, **b** $[Pt(en)_2][PtX_2(en)_2](lipid)_4$ dispersed in organic media and TEM picture (sample without staining). **c** Heat-dissocated complexes of $[Pt(en)_2](lipid)_2$ and $[PtX_2(en)_2](lipid)_2$

4.3 Supramolecular Band Gap Engineering and Solvatochromic Nanowires

The electronic structures of lipid/one-dimensional mixed valence complexes can be controlled based on the combination of metal ions [64]. On the other hand, the lipids also exert significant effect on the electronic structures. For example, the CT absorption peaks observed for $[Pt(en)_2][PtX_2(en)_2](\mathbf{19})_4$ in chlorocyclohexane (X = Cl, 2.15 eV; X = Br, 0.87 eV; X = I, less than 0.56 eV). These CT peaks are significantly red shifted compared to those observed for the crystalline perchlorates $[Pt(en)_2][PtX_2(en)_2](ClO_4)_4$ (X = Cl, 2.72 eV; X = Br, 1.95 eV) [82].

The observed red-shift is ascribed to the enhanced delocalization of the excited Pt(III)–Pt(III) states in the coordination chain, which decreases the LUMO–HOMO energy gap of the one-dimensional complex. In the lipid complexes, densely packed sulfonate groups seem to direct the electrostatically bound $Pt(en)_2$ and $PtCl_2(en)_2$ complexes to coordinate in higher density. This causes the increase in the overlap between d_z2 and p orbitals, depending on the molecular packing and chemical structure of the lipid molecules. The shortened inter-platinum (Pt^{II}–Cl–Pt^{IV}) distance would promote the excitation delocalization along the chains (Supramolecular band gap engineering) [68, 70]. The different colors observed for lipid complexes of $[Pt(en)_2][PtCl_2(en)_2]$(**17, 18, 19, 20** or **21**)$_4$ clearly indicate that the molecular information of organic molecules are reflected in the electronic structures of the linear conjugated chains.

Another unique characteristic of the lipid-complexes is solvatochromism. In the case of $[Pt(en)_2][PtCl_2(en)_2]$(**20**)$_4$, intense intervalence CT absorption was observed at 473 nm (ε: 13 846 units M^{-1} cm^{-1}) in chloroform, whereas it is observed at 513 nm (ε: 15 017 units M^{-1} cm^{-1}) in dichloromethane [69]. The observed solvatochromism is ascribed to the different degree of solvation and molecular orientation of the lipids, which significantly affected the morphology and electronic structure of these assemblies as depicted in Fig. 13. In orange chloroform dispersion, nanofibers with thickness of 20–200 nm and length of more than 20 μm were observed. On the other hand, rectangular nanocrystals (width 100 nm, length 200–300 nm) were observed in the red dichloromethane dispersion. Solvatochromism was also observed for $[Pt(en)_2][PtCl_2(en)_2]$(**19**)$_4$, in this case after re-assembly of the heat-dissociated complexes [73].

The lipid packaging of low-dimensional inorganic solid enables creation of self-assembling nanofibers. This strategy has been widely applied to inorganic complexes such as $[Mo_3Se_3]^-$ chains [84], linear iron(II) complexes [85], wheel-shaped polyoxometalate [86] and halogen-bridged platinum complexes [87]. Another class of lipophilic linear complexes are designed by introducing 1-aminoalkanes as ligands in Magnus green salt $[Pt(NH_2R)_4][PtCl_4]$ [88, 89]. Field-effect transistors comprising $[Pt(NH_2dmoc)_4][PtCl_4]$ (dmoc: (*S*)-3,7-dimethyloctyl group) as the active semiconductor layer were produced by dipping a friction-deposited transfer film of poly(tetrafluoroethylene) (PTFE) film into a super-saturated toluene solution of the Pt compound [89].

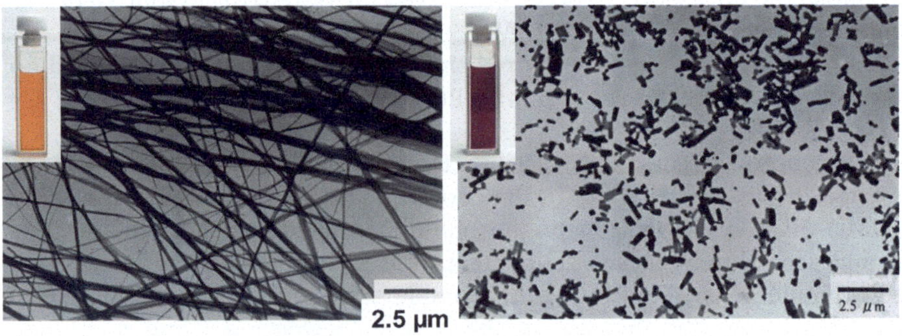

Fig. 13 Solvatochromic nanofibers formed from $[Pt(en)_2][PtCl_2(en)_2]$(**20**)$_4$. **a** In chloroform, **b** in dichloromethane. 0.6 unit mM, 15 °C [69]

4.4
Self-Assembly of Nanofibers at Interfaces

One of the important applications of the lipid (surfactant)-inorganic complexes is development of self-assembling molecular- or nano-scale electronic devices. The ability to fabricate technologically useful architectures on surfaces should be developed on the basis of a non-lithographic, self-organization process. The halogen-bridged mixed valence complexes in organic media allow the formation of thin films, which open applications of these unique nanomaterials. By spreading chloroform solutions of $[Pt(en)_2][PtCl_2(en)_2](\mathbf{18})_4$ at the air–water interface, surface monolayers are obtained [66]. To get stable monolayers which can be deposited on solid surfaces by Langmuir–Blodgett technique, it was necessary to spread the dissociated colorless complexes and to form the coordination chains at the air–water interface.

To enhance the lipophilic nature of the lipids, oleyl chains are introduced in the L-glutamate amphiphile **21** [70]. When the complex $[Pt(en)_2][PtCl_2(en)_2](\mathbf{21})_4$ is dissolved in chloroform at room temperature, the indigo color of the solid disappears, due to dissociation of the polymeric complex into the component complexes. Upon casting the colorless solution on the quartz plate, purple films were formed immediately after evaporation of the solvent. It indicates that polymerization of $[Pt(en)_2](\mathbf{21})_2$ and $[PtCl_2(en)_2](\mathbf{21})_2$ complexes occurred during the solvent evaporation. Interestingly, when the dilute chloroform solutions are dropped on carbon-coated copper TEM grid under a humid air, regular, double-layered honeycomb architecture were observed in scanning electron microscopy (Fig. 14, one side of the hexagons: ca. 650–750 nm, width of the wires: ca. 100 nm) [70]. The top honeycomb layer is connected to the basal honeycomb layer via perpendicularly oriented pillars (height: ca. 320–370 nm) at the corner of hexagons (Fig. 14, right). Self-assembling characteristics of $[Pt(en)_2][PtCl_2(en)_2](\mathbf{21})_4$ in the course of rapid solvent evaporation seem to be an essential factor for the formation of such unique stereo-nanostructures.

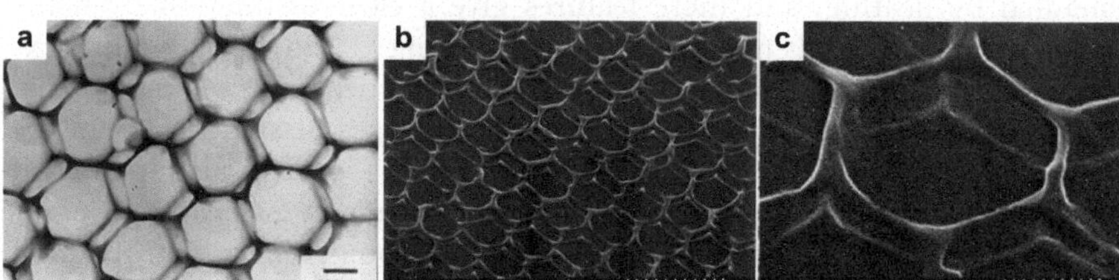

Fig. 14 TEM (**a**) and SEM (**b,c**) images of stereo-honeycomb nanoarchitectures formed from $[Pt(en)_2][PtCl_2(en)_2](\mathbf{21})_4$ [70]

Formation of two-dimensional honeycomb morphology has been observed in the solvent evaporation of copolymers [90–94], fluorinated silver nanoparticles [95] and also for carbon nanotubes [96]. In these systems, water droplets condensed from moisture on the evaporating solutions act as the template to direct formation of the honeycomb patterns. Similar mechanisms would be involved in the present system, where the rigid coordination chains formed during the solvent evaporation stabilize the stereo-honeycomb architectures. A recent report also indicates the importance of rigid polymer structures in the formation of double-layered honeycomb structure [97, 98].

4.5 Nanofibers Self-Assembled from Lipophilic Bridging Ligands and Metal Ions

1,2,4-Triazoles are known as bridging ligands and their linear iron(II) complexes have been extensively studied, because of the spin-crossover (SC) characteristics between low-spin (LS, purple, $S = 0$) and high-spin (HS, colorless, $S = 2$) configurations [99–102]. The changes in magnetism and colors are induced by external perturbation, rendering these complexes powerful candidates for information storage materials. Similar to the case of halogen-bridged mixed valence complexes, studies of the triazole complexes have been limited to the crystalline samples. Kahn et al. attempted to prepare Langmuir–Blodgett films by employing Fe^{II} complex of 4-octadecyl-1,2,4-triazole, however it resulted in deterioration of the complex at the air–water interface [101]. To develop soluble triazole complexes, we have designed a lipophilic triazole ligand **22** ($C_{12}OC_3Trz$) and synthesized their Fe^{II} and Co^{II} complexes (Fig. 15) [75, 76]. When $Co^{II}(\mathbf{22})_3Cl_2$ is dissolved in chloroform, a blue gel-like phase is formed at room temperature. Networks of nanofibers with widths of 5–30 nm are observed in atomic force microscopy. The blue color is characteristic to the tetrahedral cobalt(II) (T_d, ${}^4A_2 \rightarrow {}^4T_1(P)$). Interestingly, the blue gel-like phase turns into a pale pink solution upon *cooling*, which color indicates the formation of fragmented rod-like octahedral (O_h) complexes. The gel-like phase is reversibly generated by heating, and these features give a clear distinction from the conventional organogels that dissolve upon heating [25, 26]. This is a first example of reversible formation of heat-set gel-like networks in organic media, where the heat-set transition (from the O_h complexes in solution to the T_d complexes in gel-like phase) is enthalpically driven [76]. It is also important to note that the introduction of flexible ether-linkage in **22** promotes physical gelation and provides higher temperature responsiveness. Introduction of ether linkages in alkyl chains serves to decouple molecular orientation of different structural modules, and it often leads to superior self-assembling properties [103].

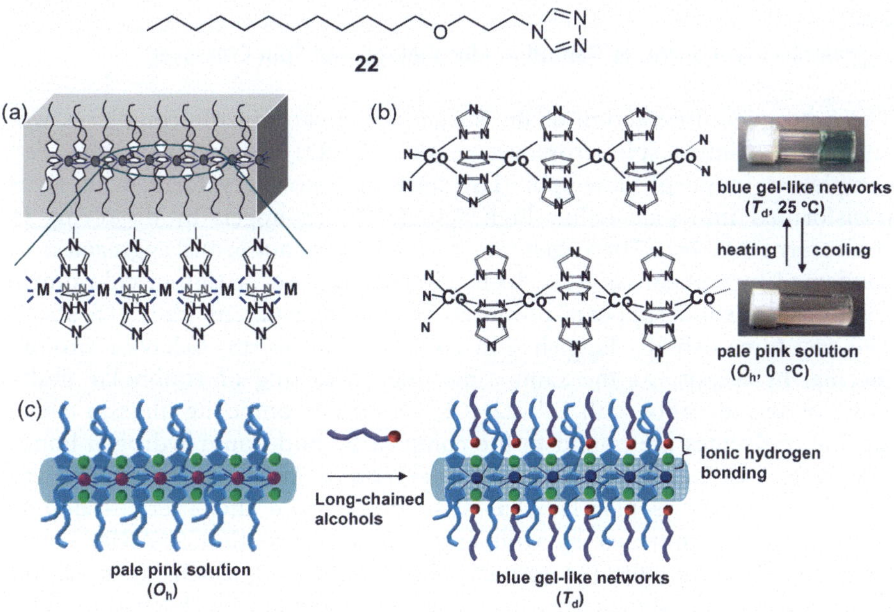

Fig. 15 Lipophilic triazole complexes. Schematic illustrations for **a** Fe(II)(**22**)$_3$Cl$_2$ and **b** thermochromic heat-set gelation observed for Co(II)(**22**)$_3$Cl$_2$ in chloroform [76]. **c** Schematic illustration of coordination structural changes induced by guest binding on the surface of Co(II)(**22**)$_3$Cl$_2$ [77]

The surface of Co^{II}(**22**)$_3$Cl$_2$ nanofibers in chloroform possesses chloride counterions, which can serve as acceptor for ionic hydrogen bonding. When long-chained alcohols were added to pink Co^{II}(**22**)$_3$Cl$_2$ solutions at 0 °C, blue organogels were obtained [77]. It occurred as a consequence of ionic hydrogen bond formation between long-chained alcohols and chloride ions (Fig. 15c), which seemingly suppressed coordination of crystal water molecules to the fragmented O_h complexes. It induced coordination structural changes from O_h to T_d and spontaneous bundling of lamellar nanofiber networks. X-ray diffraction of alcohol-doped xerogels showed increase in lamellar long spacings from 35.6 to 44.3 Å, indicating that alkyl chains adopted more perpendicular orientation with respect to the coordination main chains. That is, binding of alcohol molecules to the surface of Co^{II} (**22**)$_3$Cl$_2$ induced T_d coordination structures in which alkyl chains of ligands and alcohols are regularly aligned. Thus, the lipophilic Co^{II}(**22**)$_3$Cl$_2$ can accommodate guest alcohol molecules by changing coordination structures and orientation of alkyl chains. The binding space created on the surface of 1D complexes was referred to as "adaptable molecular clefts".

4.6 Supramolecular Control of Nanofiber Morphology and Spin Crossover

The host–guest interaction at the surface of linear triazole complexes was applied to control spin crossover of the $Fe^{II}(\mathbf{22})_3Cl_2$ complex. This Fe^{II} complex is in purple low spin (LS) state in a powdery form, while it is transformed into pale-yellow high spin (HS) organogels by dissolving in chloroform [75, 78]. These gels are formed by nanofibrous aggregates, as confirmed by transmission electron microscopy. The casting of chloroform solutions on solid supports produced purple transparent films (LS state). The cast films exhibit sluggish spin crossover (LS $\rightleftharpoons$ HS) behavior without thermal hysteresis. At the same time, the co-casting of equimolar dodecanol or tetradecanol with $Fe^{II}(\mathbf{22})_3Cl_2$ produced composite films in which alcohol molecules are bound to the complex by both ionic hydrogen bonding ($ROH\cdots Cl^-$) and van der Waals interactions. At room temperature, the cast films exhibit regular lamellar structures before and after the doping of alcohol, as confirmed by wide-angle X-ray diffraction (WAXD) measurements. Interestingly, the ternary films $Fe^{II}(\mathbf{22})_3Cl_2$/CnOH ($n = 12$, 14) showed reversible abrupt spin crossover accompanied by thermal hysteresis [78]. The observed bistability is related to dynamic structural transformations between lamellar and hexagonal structures. It provides a novel supramolecular strategy for controlling the bistability of spin crossover phenomena.

The structural dynamism of metal complex nanofibers in response to temperature changes and interactions with guest or solvent molecules prompted us to further investigate photochemical regulation of their nanostructures. A triazole ligand containing the azobenzene group formed a 1D iron(II) complex $Fe(\mathbf{23})_3(BF_4)_2$, which was soluble in common organic solvents. Atomic force microscopy (AFM) of the sample in dilute chloroform solutions showed flexible molecular wires (length: 10–100 nm, width: ca. 7 nm, Fig. 16). At the same time, $Fe(\mathbf{23})_3(BF_4)_2$ formed organogels when dispersed in chlorocyclohexane. UV-light irradiation of Fe(*trans*-**23**)$_3(BF_4)_2$ in yellow chlorocyclohexane gel (10 mM) caused fluidization into orange-colored solution of Fe(*cis*-**23**)$_3(BF_4)_2$ (Fig. 16, b → c). Visible-light illumination of the Fe(cis-**23**)$_3(BF_4)_2$ complex reversibly afforded Fe(*trans*-**23**)$_3(BF_4)_2$ complex, which is accompanied by physical gelation. Thus, the molecular wire of $Fe(\mathbf{23})_3(BF_4)_2$ shows reversible conformational changes with maintaining the linear complex structures [79].

Clérac et al. reported that the Fe^{II}(4-octadecyl-1,2,4-triazole)$_3$ complex gelatinizes dodecane and shows thermal spin-crossover [104]. Aida et al. showed that Fe^{II} complex of double-chained 1,2,4-triazole ligands display spin-crossover, which was accompanied by dissolution of the dodecane gel [105]. In these systems, however, LS complexes are maintained only in the solid state or in organogels composed of crystalline aggregates. In addition,

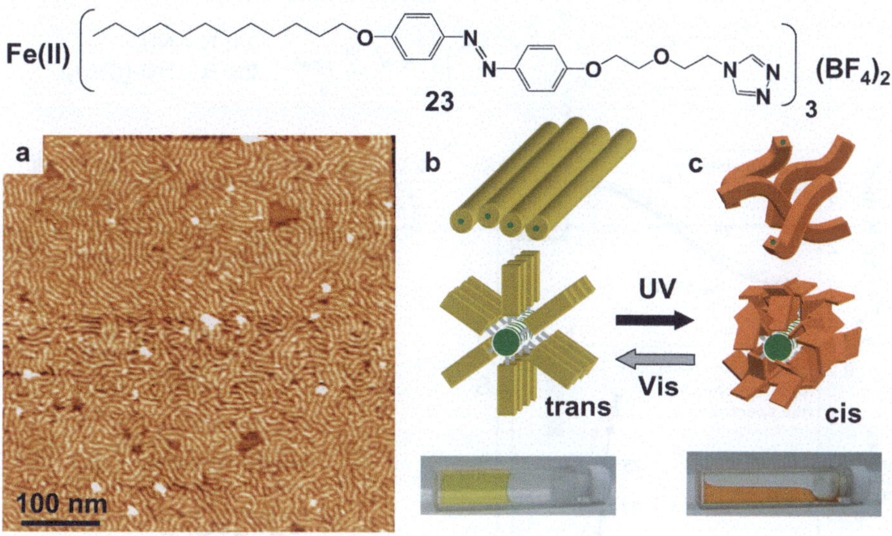

Fig. 16 **a** AFM image of Molecular wires formed in chloroform from azobenzene-containing Fe^{II}(**23**)$_3$(BF_4)$_2$. **b** Organogel formed in chlorocyclohexane (*trans*-form) and **c** after UV irradiation (*cis*-form) [79]

covalent modification of triazole ligands with bulky substituents generally causes lengthening of Fe^{II}-ligand bonds, resulting in the destabilization of LS states [106]. To address above issues, the lipid-packaging strategy was recently applied to Fe^{II}(triazole)$_3$ complexes [74]. L-Glutamate derived lipid **19** was introduced as lipophilic counteranion, and 4-amino-1,2,4-triazole (NH_2trz), 4-(2-hydroxyethyl)-1,2,4-triazole (HOC_2trz) were employed as ligands.

These complexes show sharp and reversible spin crossover (SC) in toluene (Fig. 17). The powdery lipid complexes give a spin crossover temperature (T_{SC}) of 280 K (for **24**) and 170 K (for **25**), respectively (Fig. 17a). To our surprise, these SC temperatures observed for toluene dispersions are significantly higher (by 20–100 K) than those observed in the crystalline states (Fig. 17b). The remarkable stabilization of low spin (LS) configuration in solution indicates increase in the ligand-field splitting energies, which occurs as a consequence of contraction in Fe–N and/or the nearest neighbor Fe$\cdots$Fe distances. This phenomenon is unusual, since such interactions are usually weakened in solution because of the enhanced thermal motions and interactions with solvent molecules. The unique enhancement of metal–ligand interactions in solution was ascribed to solvophobic contraction of charged coordination chains in nonpolar media [74].

AFM observation of complex **24** and **25** showed nanofibers (widths: 20–30 nm, heights: ca. 7 nm) in the LS states. In contrast, fragmented

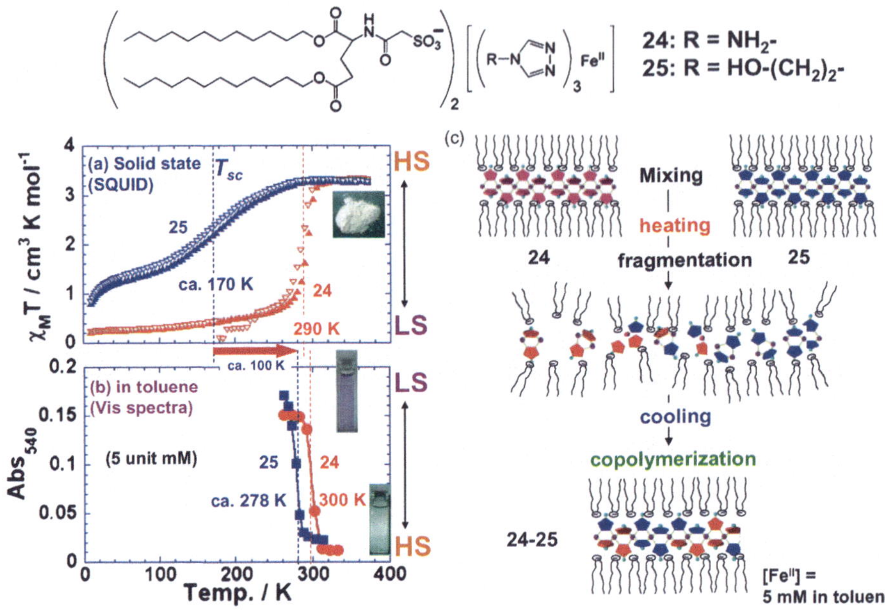

Fig. 17 **a** Spin transition in the solid state (SQUID data), **b** spin crossover in toluene dispersions (absorption intensity for LS complexes are plotted as a function of temperature), **c** supramolecular copolymerization [74]

nanofibers or dots are observed for HS complexes. These observations clearly indicate that nanofibers of LS complexes undergo thermal SC to labile HS complexes and dissociation into fragments. Lipids with enhanced thermal mobility would render the fragments to adopt a HS state, as indicated by the abrupt spin state changes. The observations of developed nanofibers after cooling the heat-dispersed samples and the recovery of LS complexes clearly indicated reversible self-assembly of HS fragments into the original, LS nanofibers. The reversible self-assembly was confirmed by copolymerization of **24** and **25** after the heat treatment (Fig. 17c). As spin crossover premises an unchanged coordination number at central atoms, the changes in spin state involving disintegration of coordination bonds would be better described as *spin conversion directed by self-assembly.*

The lipid-packaging of charged coordination chains not only provides solubility to the linear metal complexes, but also improves their potential functions. The solvophobic contraction of charged metal complexes in organic media is a new concept, which would be widely applicable to developing functional nanofibers that show improved performances compared to those in the solid state. It would push back the boundaries of existing coordination polymer chemistry.

5 Conclusion and Outlook

Self-assembly offers an attractive strategy to construct functional nanofibers. Amphiphilic self-assembly of components that belong to different molecular families or disciplines allows development of unique properties that have not been created in the framework of conventional studies. The solvophilic–solvophobic interactions play a pivotal role in the determination of mesoscopic-scale supramolecular architectures, and this is a feature not appreciated in classical supramolecular chemistry. Since many supramolecular nanofibers have been developed, one of the next issues would obviously be finding functions unique to their nanoarchitectures. Introduction of biological molecules as essential components for nanofiber self-assembly has been shown to provide bridges that link materials science and in vivo applications.

From a chemical point of view, it is interesting to create new properties which are based on the interplay of components self-organized in nanoarchitectures. The generation of unique electronic structures based on self-assembly (Supramolecular band gap engineering), the heat-set gel-like networks in organic media, and unusual stabilization of low spin states in organized solutions are examples of surprise results that were not previously found. The growth of electronically conjugated networks is reminiscent of neural development, and we envisage that such nanofibers would be applicable to the design of neuromorphic nanodevices, which display growth and self-restoration in response to chemical or electronic stimuli. The combination of bottom-up self-assembly and top-down techniques will also provide an opportunity for design of molecular systems, where elements of molecular information, their transmission and translation, amplification and feedback are integrated in supramolecular circuits to achieve synergistic functions.

Acknowledgements This work was financially supported by a Grant-in-Aid for Scientific Research (A) (No. 19205030), a Grant-in-Aid for the Global COE Program, "Science for Future Molecular Systems" from the Ministry of Education, Culture, Sports, Science and Technology of Japan, and by JST CREST.

References

1. Klug A (1983) Angew Chem Int Ed 22:565
2. Alberts B, Bray D, Lewis J, Raff M, Roberts K, Watson JD (1994) Molecular Biology of the Cell, 3rd edn. Garland Publishing Inc, New York
3. Lehn J-M (2002) Proc Natl Acad Sci USA 99:4763
4. Kunitake T (1992) Angew Chem Int Ed 31:709
5. Fuhrhop JH, Köning J (1994) Membranes and molecular assemblies: the synkinetic approach. The Royal Society of Chemistry, Cambridge
6. Fuhrhop JH, Wang T (2004) Chem Rev 104:2901

7. Shimizu T, Iwaura R, Masuda M, Hanada T, Yase K (2001) J Am Chem Soc 123:5947
8. Kimizuka N (2002) In: Yui N (ed) Supramolecular Architectures toward biological applications: Supramolecular Design for Biological Applications, Chapt 18. CRC press, Boca Raton, p 373
9. Kimizuka N (2003) Curr Opin Chem Biol 7:702
10. Kimizuka N (2005) In: Ciferri A (ed) Soluble Amphiphilic Nanostructures and Potential Applications: Supramolecular Polymers, 2nd ed, Chap 13. Taylor & Francis, Boca Raton, p 481
11. Hartgerink JD, Beniash E, Stupp SI (2002) PNAS 99:5133
12. Hartgerink JD, Beniash E, Stupp SI (2001) Science 294:1684
13. Stendahl JC, Rao MS, Guler MO, Stupp SI (2006) Adv Funct Mater 16:499
14. Paramonov SE, Jun HW, Hartgerink JD (2006) J Am Chem Soc 128:7291
15. Marini DM, Hwang W, Lauffenburger DA, Zhang S, Kamm RD (2002) Nano Lett 2:295
16. Zhang S (2003) Nat Biotechnol 21:1171
17. Hamley IW (2007) Angew Chem Int Ed 46:8128
18. Cornelissen JJLM, Fischer M, Sommerdijk NAJM, Nolte RJM (1998) Science 280:1427
19. Nowak AP, Breedveld V, Pakstis L, Ozbas B, Pine DJ, Pochan D, Deming TJ (2002) Nature 417:424
20. Cui H, Chen Z, Zhong S, Wooley KL, Pochan DJ (2007) Science 317:647
21. Bae J, Choi JH, Yoo YS, Oh NK, Kim BS, Lee M (2005) J Am Chem Soc 127:9688
22. Ryu JH, Lee E, Lim Y, Lee M (2007) J Am Chem Soc 129:4808
23. Moon KS, Kim HJ, Lee E, Lee M (2007) Angew Chem Int Ed 46:6807
24. Estroff LA, Hamilton AD (2004) Chem Rev 104:1201
25. George M, Weiss RG (2006) Acc Chem Res 39:489
26. Ishi-i T, Shinkai S (2005) Top Curr Chem 258:119
27. Brunsveld L, Folmer BJB, Meijer EW, Sijbesma RP (2001) Chem Rev 101:4071
28. Kaler EW, Murthy AK, Rodriguez B, Zasadzinski JA (1989) Science 245:1371
29. Zemb T, Dubois M, Demé B, Gulik-Krzywicki T (1999) Science 283:816
30. Muñoz S, Gokel GW (1996) Inorg Chim Acta 250:59
31. Kunitake T, Ishikawa Y, Shimomura M, Okawa H (1986) J Am Chem Soc 108:327
32. MacDonald JC, Whitesides GM (1994) Chem Rev 94:2383
33. Prins LJ, Reinhoudt DN, Timmerman P (2001) Angew Chem Int Ed 40:2382
34. Hoeben FJM, Jonkheijm P, Meijer EW, Schenning APHJ (2005) Chem Rev 105:1491
35. Bouteiller L (2007) Adv Polym Sci 207:79
36. Nowick JS, Chen JS, Noronha (1993) J Am Chem Soc 115:7636
37. Kawasaki T, Tokuhiro M, Kimizuka N, Kunitake T (2001) J Am Chem Soc 123:6792
38. Nakashima N, Asakuma S, Kunitake T (1985) J Am Chem Soc 107:509
39. Iwaura R, Yoshida K, Masuda M, O-Kameyama M, Yoshida M, Shimizu T (2003) Angew Chem Int Ed 42:1009
40. Zhang Y, Yang Z, Yuan F, Gu H, Gao P, Xu B (2004) J Am Chem Soc 126:15028
41. Oda R, Huc I, Candau SJ (1988) Angew Chem Int Ed 37:2689
42. Aimé C, Manet S, Satoh T, Ihara H, Park KY, Godde F, Oda R (2007) Langmuir 23:12875
43. Nakashima T, Kimizuka N (2002) Adv Mater 14:1113
44. Ringler P, Schulz GE (2003) Science 302:106
45. Seeman NC (1997) Acc Chem Res 30:357
46. Sancenón F, Descalzo AB, Martínez M, Mianda MA, Soto J (2001) Angew Chem Int Ed 40:2640
47. Buterfield M, Waters ML (2003) J Am Chem Soc 125:9580
48. Sazani PL, Larralde R, Szostak JW (2004) J Am Chem Soc 126:8370
49. Morikawa MA, Yoshihara M, Endo T, Kimizuka N (2005) J Am Chem Soc 127:1358

50. Shiraki T, Morikawa MA, Kimizuka N (2008) Angew Chem Int Ed 47:106
51. Faul CFJ, Antonietti M (2003) Adv Mater 15:673
52. Faul CFJ, Antonietti M (2002) Chem Eur J 8:2764
53. Guan Y, Antonietti M (2002) Langmuir 18:5939
54. Guan Y, Yu SH, Antonietti M, Böttcher C, Faul CFJ (2005) Chem Eur J 11:1305
55. Wang Z, Ho KJ, Medforth CJ, Shelnutt JA (2006) Adv Mater 18:2557
56. Shenning APHJ, Meijer EW (2005) Chem Commun, p 3245
57. Mathias JP, Simanek EE, Whitesides GM (1994) J Am Chem Soc 116:4326
58. Mobian P, Kern JM, Sauvage JP (2003) J Am Chem Soc 125:2016
59. Hansenknopf B, Lehn J-M, Boumediene N, Gervais AD, Dorsselaer AV, Kneisel B, Fenske D (1997) J Am Chem Soc 119:10956
60. Fujita M, Tominaga M, Hori A, Therrien B (2005) Acc Chem Res 38:369
61. Ruben M, Breuning E, Lehn J-M, Ksenofontov V, Renz F, Gütlich P, Vaughan GBM (2003) Chem Eur J 9:4422
62. Gatteschi D (1994) Adv Mater 6:635
63. Kimizuka N, Oda N, Kunitake T (1998) Chem Lett 27:695
64. Kimizuka N, Lee SH, Kunitake T (2000) Angew Chem Int Ed 39:389
65. Kimizuka N, Oda N, Kunitake T (2000) Inorg Chem 12:2684
66. Kimizuka N, Yamada K, Kunitake T (2000) Mol Cryst Liq Cryst 342:103
67. Kimizuka N (2000) Adv Mater 12:1461
68. Lee SH, Hatanaka Y, Kimizuka N (2002) J Nanosci 5/6:391
69. Lee SH, Kimizuka N (2002) Chem Lett 31:1252
70. Lee SH, Kimizuka N (2002) Proc Natl Acad Sci USA 99:4922
71. Kimizuka N, Hatanaka Y, Kunitake T (2003) Supramolecular Assemblies Consisting of Naphthalene-containing Anionic Amphiphiles and One-dimensional Halogen-bridged Platinum Complexes. In: Geckeler K (ed) Advanced Macromolecular and Supramolecular Materials and Processes. Kluwer Academic/Plenum Publishers, London, p 109
72. Yasui K, Kimizuka N (2005) Chem Lett 34:248
73. Kuroiwa K, Oda N, Kimizuka N (2006) Sci Technol Adv Mater 7:629
74. Matsukizono H, Kuroiwa K, Kimizuka N (2008) J Am Chem Soc 130:5622
75. Kimizuka N, Shibata T (2000) Polym Prepr Japan 49:3774
76. Kuroiwa K, Shibata T, Takada A, Nemoto N, Kimizuka N (2004) J Am Chem Soc 126:2016
77. Kuroiwa K, Kimizuka N (2008) Chem Lett 37:192
78. Kuroiwa K, Shibata T, Sasaki S, Ohba M, Takahara A, Kunitake T, Kimizuka N (2006) J Polym Sci A 44:5192
79. Kume S, Kuroiwa K, Kimizuka N (2006) Chem Commun, p 2442
80. Matsukizono H, Kuroiwa K, Kimizuka N (2008) Chem Lett 37:446
81. Okamoto H, Yamashita M (1988) Bull Chem Soc Japan 71:2023
82. Wada Y, Mitani T, Yamashita M, Koda T (1985) J Phys Soc Japan 54:3143
83. Kishida H, Matsuzaki M, Okamoto H, Manabe T, Yamashita M, Taguchi Y, Tokura Y (2000) Nature 405:929
84. Messer B, Song JH, Huang M, Wu Y, Kim F, Yang P (2000) Adv Mater 12:1526
85. Kurth DG, Meister A, Thünemann AF, Förster G (2003) Langmuir 19:4055
86. Polarz S, Smarsly B, Antonietti M (2001) Chem Phys Chem 2:457
87. Matsushita N, Taira A (1999) Synth Met 102:1787
88. Bremi J, Brovelli D, Caseri W, Hähner G, Smith P, Tervoort T (1999) Chem Mater 11:977

89. Caseri WR, Chanzy HD, Feldman K, Fontana M, Smith P, Tervoort TA, Goossens JGP, Meijer EW, Schenning APHJ, Dolbnya IP, Debije MG, De Haas MP, Warman JM, Van de Craats AM, Friend RH, Sirringhaus H, Stutzmann N (2003) Adv Mater 15:125
90. Widawski G, Rawiso M, François B (1994) Nature 369:387
91. François B, Pitois O, François J (1995) Adv Mater 7:1041
92. Karthaus O, Maruyama N, Cieren X, Shimomura M, Hasegawa H, Hashimoto (2000) Langmuir 16:6071
93. Shimomura M. In: Chiferri A (ed) Supramolecular Polymers. Marcel Dekker, New York, p 471
94. Govor LV, Bashmakov LA, Kaputski FN, Pientka M, Parisi J (2000) Macromol Chem Phys 201:2721
95. Yonezawa T, Onoue S, Kimizuka N (2001) Adv Mater 13:140
96. Takamori H, Fujigaya T, Yamaguchi Y, Nakashima N (2007) Adv Mater 19:2535
97. Yabu H, Tanaka M, Ijiro K, Shimomura M (2003) Langmuir 19:6297
98. Mahajan SS, Paranji R, Mehta R, Lyon RP, Atkins WM (2006) Bioconjug Chem 16:1019
99. Kahn O, Martinez CJ (1988) Science 279:44
100. Kröber J, Codjovi E, Kahn O, Grolière F (1993) J Am Chem Soc 115:9810
101. Armand F, Badoux C, Bonville P, Teixier AR, Kahn O (1995) Langmuir 11:3467
102. Kahn O (2000) Acc Chem Res 33:647
103. Kimizuka N, Kawasaki T, Kunitake T (1993) J Am Chem Soc 115:4387
104. Roubeau O, Colin A, Schmitt V, Clérac R (2004) Angew Chem Int Ed 43:3283
105. Fujigaya T, Jiang DL, Aida T (2007) Chem Asian J 2:106
106. Fujigaya T, Jiang DL, Aida T (2005) J Am Chem Soc 127:5484

Adv Polym Sci (2008) 219: 27–68
DOI 10.1007/12_2008_169

Published online: 16 July 2008

Self-Assembled Peptide Nanofibers

Nobuyuki Higashi (✉) · Tomoyuki Koga

Department of Molecular Chemistry & Biochemistry, Faculty of Science & Engineering, Doshisha University, Kyotanabe, Kyoto 610-0321, Japan
nhigashi@mail.doshisha.ac.jp

1 **Introduction** 28

2 **β-Sheet-Based Self-Assembled Nanofibers** 29
2.1 Nanofiber Formation from Artificial Peptides and Peptidomimetics Through Self-Assembly 30
2.2 Stimuli-Responsive Self-Assembling System into Nanofibers 41
2.3 Hybridization of Peptide Nanofibers and Synthetic Polymers 46
2.4 Applications in Biomedical Engineerings 52
2.5 Applications for Nano-Templates 53

3 **Helix-Based Self-Assembled Nanofibers** 56
3.1 Coiled-Coil Motif 56
3.2 Another Class of Helix Nanofibers 62

4 **Summary** 65

References 66

Abstract Molecular self-assembly is a powerful approach being explored for novel supramolecular nanostructures and bio-inspired nanomaterials. In this article, we focus on recent research concerning the self-assembly of de novo designed artificial peptides and peptidomimetics into nanofiber structures, specifically towards developing a new class of soft-materials. These nanofiber architectures have potential use not only in biomedical applications, such as 3D-matrix scaffolds for tissue engineering and biomineralization, but also in nanotechnology such as nano-templates and dimension-regulated functional nano-objects.

Keywords Artificial peptides · Coiled-coil structure · Helix structure · Nanobiotechnology · Nanofiber · Self-assembly · β-Sheet structure

Abbreviations

Amino Acids:
Ala Alanine (A)
Arg Arginine (R)
Asn Asparagine (N)
Asp Aspartic acid (D)
Cys Cysteine (C)
Gln Glutamine (Q)
Glu Glutamic acid (E)

Gly Glycine (G)
His Histidine (H)
Ile Isoleucine (I)
Leu Leucine (L)
Lys Lysine (K)
Met Methionine (M)
Phe Phenylalanine (F)
Pro Proline (P)
Ser Serine (S)
Thr Threonine (T)
Trp Tryptophan (W)
Tyr Tyrosine (Y)
Val Valine (V)

1 Introduction

A number of natural biomaterials exhibit intelligent functions and extraordinary properties, at which various biopolymers (such as proteins, polynucleotides, and polysaccharides) assemble appropriately at nano-scale and construct structural frames based on their optimized and rational molecular design. In nature, such structuring processes by biopolymers are usually guided via self-assembly, and those components work separately and/or cooperatively with each other. Progress in the design and characterization of such self-organization principles and growing interest in nano-scale molecular engineering provide strong incentives not only to fabricate novel nanostructured materials but also to our understanding of such bio-related macromolecular assembly, including protein folding. In this context, several molecular assembling systems based on peptides, proteins, DNA, and RNA have been explored. Our focus has been on peptide-based nano-assemblies. Peptides and proteins are especially versatile building units for this purpose, because they have ability to hierarchically self-organize into precisely defined three-dimensional (3D)-nanostructures with a much higher level of regularity than those which can be produced from conventional synthetic block copolymers.

For example, α-helix and β-sheet are the major secondary structural motifs with well-regulated geometry organizing the 3D-structure of proteins. In the α-helix, the peptide bond planes are parallel to the axis of the helix. In this geometry, each peptide forms two intra-molecular hydrogen bonds, one to the peptide bond of the fourth residue above and the second to the peptide bond of the fourth amino acid below in the primary structure. The side chains in an α-helix are on the outside of the spiral structure. On the other hand, in the β-sheet, a polypeptide chain is hydrogen bonded to another polypeptide chain aligned in a parallel or an antiparallel direction (inter-molecular hydrogen bond). Hydrogen bonded β-strands appear like a pleated sheet, and the side chains project above and below the sheet-like structure. It is generally considered that the sequence of amino acids is important in determining the properties of the protein secondary structure. The protein high-order structure formed by the specific interaction among such secondary structural elements plays a key role in defining their functions and natures. In other words, transferring such protein self-organization systems at the molecular level by the use of artificial peptides and peptidomimetics with appropriate conformations towards the fabrication of synthetic nanostructured materials would be of great interest. This approach enables the construction of 3D-structures with diverse shapes, sizes and functions and will complement the limit of a conventional top-down lithographic approach.

In this paper, we review recent research concerning the controlled self-assembly of artificial peptides and peptidomimetics into nanofiber architectures, and their potential applications in the field of biotechnology and nanotechnology. This field is undergoing a rapid growth from the viewpoint of both materials science and biological studies aimed at understanding the protein aggregation. Numerous molecules, such as native proteins, synthetic peptides, and lipid-like amino acid(short peptide)-based amphiphiles, have been reported to self-assemble into one-dimensional fibrous nano-objects. In particular, since the it is easy to prepare, functionalize, and manipulate synthetic peptides' self-assembling characteristics by changing the amino acid sequence, artificial β-strand and α-helical building units have mainly been employed as structural elements to fabricate the organized nanofiber-architectures. Therefore, this paper focus mainly on the (1) β-sheet-based and (2) α-helix-based self-assembled nanofibers from de novo designed synthetic peptides, specifically towards developing a new class of soft-materials.

2
β-Sheet-Based Self-Assembled Nanofibers

The self-assembly of peptides and proteins into β-sheet-rich fibrillar structures has attracted much attention as a powerful approach for designing novel functional nanomaterials, and because of their association with neurodegenerative diseases, such as Alzheimer's disease [1–6]. Several proteins were identified in amyloid aggregates [7, 8], and these aggregates were found to have a similar core structure (a straight, long, cross-β structure) [9]. Non-pathogenic proteins and synthetic peptides also form amyloid-like fibrils that structurally resemble in vivo fibrils [10–12]. Thus, the ability to form amyloid-like fibrils from a wide range of synthetic peptides gives access to a large number of model systems with which to study the process of fibril formation in more detail. On the other hand, from the standpoint of materials science, the β-sheet motifs are attractive as a building block in the design of supramolecular nanofibers with well-defined structures. Such unique structures are expected to have potential as novel nano-scaled soft-materials with diverse applications in the fields of nanotechnology and biotechnology.

In this section, we focus on the research of β-sheet-based self-assembled nanofibers from de novo designed artificial peptides and peptidomimetics aimed for applying in materials science.

2.1 Nanofiber Formation from Artificial Peptides and Peptidomimetics Through Self-Assembly

Three main types of artificial peptide motifs have been developed to construct β-sheet-based nanofibers through self-assembly, as shown in Fig. 1. These peptides and peptidomimetics have hydrophobic and hydrophilic segments as a common feature. The hydrophobic segment seems to provide an important driving force for self-assembly, as well as hydrogen bonding and electrostatic interactions, especially in aqueous media.

The first type of self-assembling building units are synthetic poly(oligo)peptides, which consist of natural amino acids with hydrophobic and hydrophilic characters (type I in Fig. 1). The self-assembling properties and the resultant nanostructures, including the physical property of these building units, are strongly affected by amino acid sequence, stereochemistry, molecular size, and environmental conditions such as pH and ionic strength.

We have first prepared block-type simple and amphiphilic peptides composed of 16-residue Leu and Lys as self-assembling building units (amino acid sequences: LLLLKKKKKKKKLLLL (LKL16), KKKKKKKKKKKKLLLL (LK16), and KKKKKKLLLLKKKKKK (KLK16)) [13]. The LKL16, LK16, and KLK16 were designed to have a tetra-Leu domain, which provided the hydrophobic driving force for self-assembly, and was located at either both ends (N- and C-termini), at one end (C-terminus), or the center of the peptide, respectively. These peptides should clarify the effect of the arrangement of the hydrophobic domain in the peptide chain on their self-assembling properties in water. In addition, we anticipated that it might be possible to control the distribution on the high-order structures by manipulating solution pH, owing to the charged condition of the Lys residues.

For LK16 and KLK16 (40 μM), the CD spectra gave a typical pattern of right-handed α-helical peptides with two negative maxima, one at 222 nm and one at 204 nm in the high pH region at 10.6. By lowering the pH value to 5.8, both peptides of LK16 and KLK16 formed random coil structures. These conformations were time-independent within the period of 1 week. On the other hand, conformational property of LKL16 differed considerably from those of LK16 and KLK16. Although the LKL16 also existed as a random coil structure at pH 5.8, the structural transition into β-sheet were observed by elevating the pH value up to 9.4. At above pH 9.5, the CD spectra showed

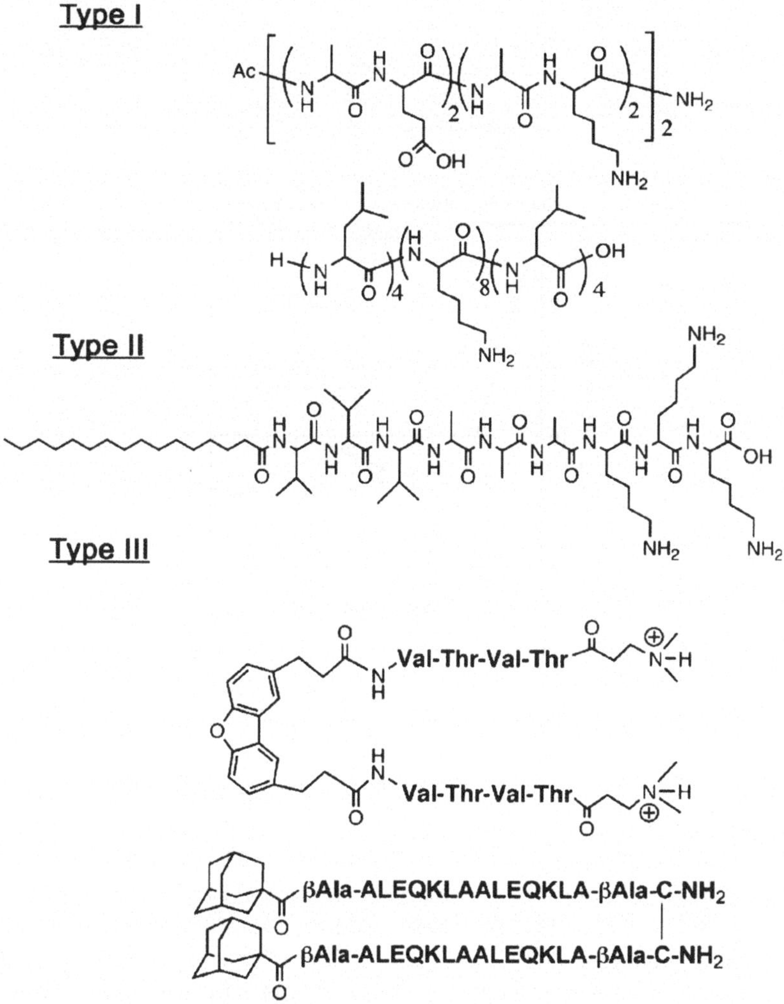

Fig. 1 Chemical structures of various types of artificial peptides and peptidomimetics (type I–III) that self-assemble into β-sheet-based nanofibers

the aggregated β-sheet structure. Interestingly, in the case of LKL16 (40 μM), spontaneous α(random)-to-β conformational transition with incubation time were observed in the narrow pH range of 8.8–9.4 (Fig. 2B). The FTIR spectrum of the LKL16 obtained after 24 h-incubation at pH 9.2 supported the CD results and suggested that the β-structures are classified as antiparallel β-strands on the basis of the absorptions at 1687 and 1625 cm^{-1}. Depending on the variety of conformational properties, these designed peptides showed the diverse self-assembly. Figure 3 shows the self-assembly of LKL16 associated with the β-sheet formation at pH 9.2. The AFM image obtained at 3 min re-

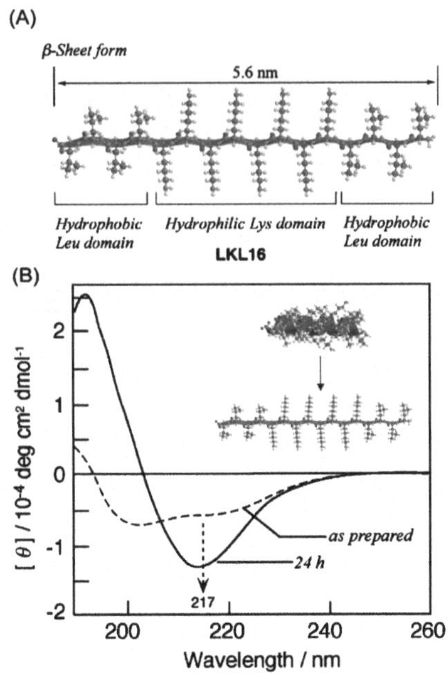

Fig. 2 **A** Molecular structure of the amphiphilic triblock-type oligopeptide, LKL16, used as a building unit for self-assembly. **B** CD spectral change of LKL16 in Tris-buffer at pH 9.2. The peptide was incubated at room temperature for the time indicated. [LKL16] = 40 μM

vealed the presence of only globular aggregates (Fig. 3A). The average height of the globular species was determined to be 0.8 ± 0.3 nm, indicating that the LKL16 exists in aqueous solution as a monomer. On the other hand, well-developed peptide nanofibers were observed instead of globular species after incubation for 5–24 h (Fig. 3B). The nanofibers mainly possessed a diameter of ca. 6.0 nm and a clearly visible left-handed twist that repeats along the fiber length (Fig. 4, left image). The periodicity of this repeat was ca. 50 nm, and the fiber height oscillated between 5.5 nm and 6.5 nm. The observed twist in LKL16-nanofiber and its handedness are probably due to the right-handed twist along the backbone of a β-strand made of L-amino acids [14]. Similarly, left-handed β-sheet assemblies (helical ribbons, tapes, fibrils) have also been observed from synthetic amphiphilic L-oligopeptides [15–17, 26, 32]. It is considered that the packing of right-handed β-strands a causes left-handed spiral in the β-sheet [18]. In fact, the nanofiber formed by LKL16(D), which consisted of D-amino acids, showed identical periodicity and were also clearly twisted, but the twist direction was right-handed (Fig. 4, right image) [19]. In other words, this result indicates the possibility that the helicity of self-assembled nanofibers can be controlled by the chirality of the amino acids in the constituent peptide species, although some peptides have been reported to form non-twisted β-sheet nanofibers [20, 31]. The morphological

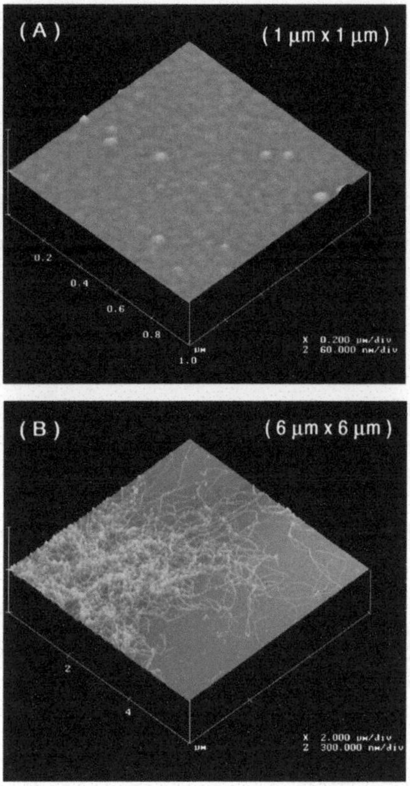

Fig. 3 Tapping-mode AFM images of LKL16 obtained after incubation for 3 min (**A**) and 24 h (**B**) at pH 9.2. [LKL16] = 40 µM. *z*-Scales: 60 nm (**A**) and 300 nm (**B**)

property of LKL16 was strongly affected by solution pH. At a lower pH (7.0) region, no aggregation was observed, indicating that the LKL16 is predominantly in the monomeric random coil conformation. On the other hand, at higher pH of 10.6, the peptide was assembled rapidly and formed poorly organized amorphous aggregates. Such rapid aggregation of LKL16 seems to restrict mobility of the peptide chains, resulting in prevention of appropriate packing of peptide chains necessary for further organization. Thus, it seems that the development of regular intermolecular interactions involving extended regions of peptides are important for the formation of highly organized nanofibers. At ca. pH 9, the Lys residues of LKL16 are partly protonated, and this charge will probably provide sufficient intermolecular repulsion to permit controlled growth of nanofibers as well as sufficient mobility of peptide chains for appropriate packing. Note that this self-assembly of LKL16 into nanofibers is reversible by pH adjustment, namely the nanofiber is decomposed into monomeric random coil by lowering the solution pH. Self-fabricated nanostructure could also be controlled by tuning the peptide

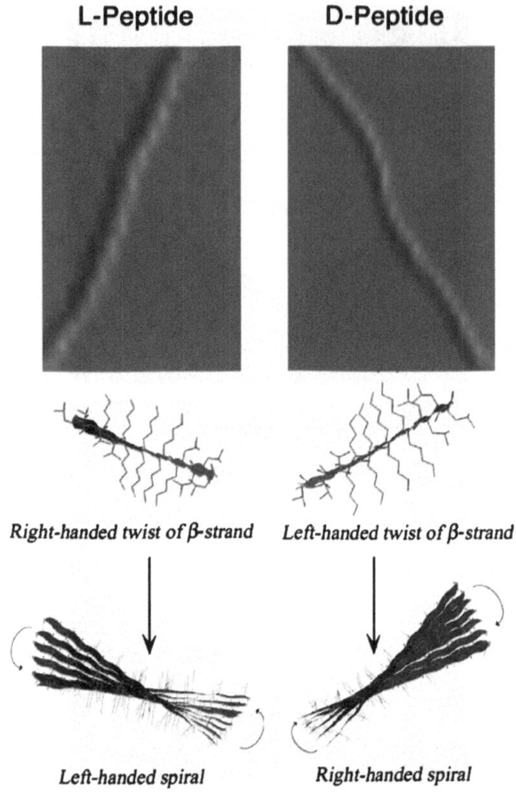

Fig. 4 Amplitude AFM images of LKL16-nanofiber, LKL16(D)-nanofiber, and their plausible models

sequence, namely the arrangement of hydrophobic domain in the peptide chain. It was found that spherical aggregates with ca. 3.0–5.0 nm heights were formed from LK16 (Fig. 5). The measured dimensions of these aggregates are in good agreement with the twice value of the molecular length of α-helical LK16 (ca. 2.4 nm), indicating the formation of micelle-like structures consisting of a hydrophobic Leu domain core and the shell of a Lys domain (see Fig. 5). The peptide LK16 has the tetra-Leu domain at only one chain terminus, and this diblock-type structure will probably stabilize such micelle-structures. On the other hand, the α-helical KLK16, in which the tetra-Leu domain was arranged at the center, self-assembled into plate-shape aggregates with height of around 1.5 nm and lengths of 100–500 nm. The observed heights of the plates correspond to the diameter of a polylysine helix (ca. 1.4 nm). Therefore, these plate-shaped aggregates probably consist of a helix monolayer, in which KLK16-helices interact with each other in two-dimensions through the hydrophobic interaction between Leu residues located at the center, as well as the helix–macrodipole interaction [21]. Thus,

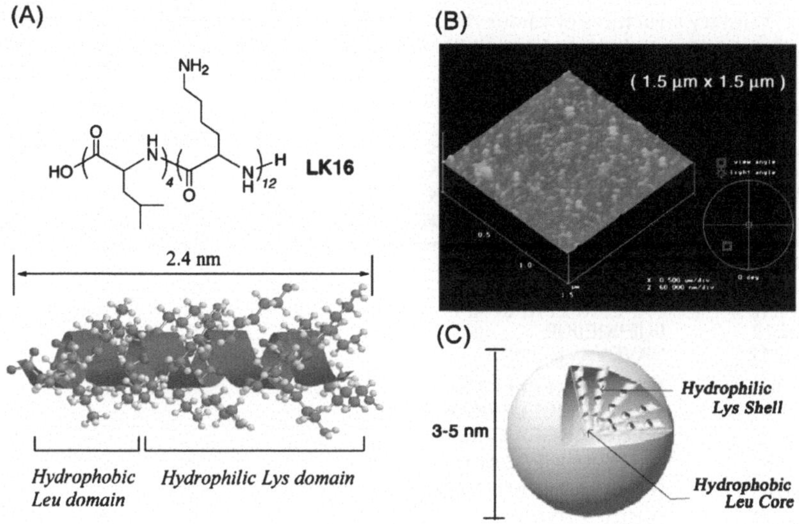

Fig. 5 **A** Chemical structure of the amphiphilic diblock-type oligopeptide, LK16, and its molecular dimension in α-helical form. **B** Tapping-mode AFM image of the self-assembled nanoparticle obtained from LK16. **C** Proposed model of the nanoparticle from α-helical LK16

by tuning the peptide sequence (e.g., location of hydrophobic domains) and environmental conditions, such as solution pH and reaction time, a wide variety of self-organized 3D-architectures with various nanostructures were obtained by these self-assembling systems.

Most typical peptide sequences for β-sheet formation in type I series alternate placement of hydrophobic and polar amino acids. These sequences provide two distinctive faces: one hydrophobic and the other hydrophilic in β-strands. The hydrophobic faces shield themselves from water thus facilitating their self-assembly in water. The hydrophilic face forms the outside of the β-sheet, which can be designed to interact with each other through complementary electrostatic interactions. A number of β-sheet-forming peptides with alternating hydrophobic and hydrophilic patterns were prepared, and their self-assembling behaviors were examined in the view of the hydrophobic side chains, charged side chains, and molecular weight, etc. [20, 22–32]. These peptides not only self-assemble to form stable nanofibers, but also form high-order nanofiber-scaffolds, namely hydrogels. Table 1 listed a part of this series of peptides. This class of oligopeptides was originally discovered by Zhang et al. [22]. They designed 16-mers alternating peptide, made up of hydrophobic (Ala) and hydrophilic (Glu and Lys) amino acids, Ac-AEAEAKAKAEAEAKAK-CONH$_2$ (EAK16-II), and demonstrated that the EAK16-II self-assembled into stable β-sheet nanofiber 3D-matrix with $\sim$10-20 nm in diameter in aqueous solution under the presence of monovalent metal ions, such as Li^+ and Na^+. The critical aggregation concentration of

Table 1 Primary structures of vaious β-sheet nanofiber-forming peptides with alternating sequence of hydrophobic and hydrophilic amino acids

Peptides	*Amino Acid Sequences*
EAK16-II	AEAEAKAKAEAEAKAK
EAK16-IV	AEAEAEAEAKAKAKAK
EAK16-I	AEAKAEAKAEAKAEAK
poly(EAK)$_9$	(AEAEAKAKAEAEAKAK)$_9$
KFE8	FKFEFKFE
KFE12	FKFEFKFEFKFE
KFE16	FKFEFKFEFKFEFKFE
KIE12	IKIEIKIEIKIE
KVE12	VKVEVKVEVKVE
00	FEFEFKFKFEFEFKFK
81	GEAEVKLKIELEVKAK
82a	IELEVKAKGEAEVKLK
83	IEAEGKGKIEGEAKIK
RADA16	RADARADARADARADA
MAX1	(VK)$_4$V^{D}PPT(KV)$_4$
MAX1(L)	(VK)$_4$VPPT(KV)$_4$
FF	PKFKFFEFEP
FI	PKFKIIEFEP
VI	PKVKIIEVEP

the EAK16-II was estimated to be $\sim$ 0.1 mg/ml by surface tension measurements [23]. Zhang and Chen et al. also investigated the effect of the charge distribution of EAK16 on the nanostructure formation [24, 25]. For this purpose, three types of EAK16 were studied: EAK16-I, -II, and -IV, which have the same amino acid composition, but have the different charge distribution (−−++−−++ for EAK16-II, −−−−++++ for EAK16-IV, and −+−+−+−+ for EAK16-I). AFM measurements showed that EAK16-I and -II formed nanofiber assemblies regardless of pH value, whereas EAK16-IV formed globular structures at neutral condition of pH between 6.5 and 7.5. When the pH value was lowered below 6.5 or raised above 7.5, a transition from globular to fibrillar structures was observed for EAK16-IV. From the FTIR and theoretical studies, they showed that the hairpin structure is energetically stable for EAK16-IV, but the chain entropy of EAK16-I and -II favors relatively stretched conformation. Based on these results, they concluded that individual EAK16-IV peptides fold first into β-hairpins, exposing their hydrophobic faces to water, and these hairpins then aggregate into globular structures by hydrophobic interactions. In the class of one sequence of the family, named KFE, phenylalanines are employed as hydrophobic side chains instead of alanine. These peptides self-assemble into left-handed helical ribbons [26, 27], and also form a β-sheet-based 3D-matrix that behave as an elastic gel upon addition of a sufficient concentraion of salt. Lauffenburger et al. examined the effect of the number of repeats of KFE sequence on the β-sheet-based 3D-matrix formation by using KFE8, KFE12, and KFE16 at pH 3 [28]. From a rheometric study, they determined critical coagulation concentration (CCC) of NaCl salt required to 3D-matrix (gel) formation. The CCC values were estimated to be $1.0 \sim 5.0$ mM NaCl for KFE8, $0.1 \sim 1.0$ mM NaCl for KFE12, and $1.0 \sim 5.0$ mM NaCl for KFE16 at pH 3, meaning the biphasic trend was observed as oligopeptide length was increased. The authors explained that such a biphasic trend was due to an unequal competition between a greater hydrophobic (favorable) effect and a greater entropic (unfavorable) effect as the peptide length was increased. Muller et al. successfully prepared the polymeric EAK16-II, poly(EAK)$_9$, by recombinant DNA method, and also compared the self-assembling behavior of high-molecular weight peptides with that of low molecular weight peptides (EAK16-II) [29]. The poly(EAK)$_9$ was demonstrated to form well-defined antiparallel β-sheet nanofibers with 10–20 nm diameter, that structurally resembled that from EAK16-II. On the other hand, the poly(EAK)$_9$-based hydrogels obtained under physiological condition (10 mM sodium phosphate buffer at pH 7) was significantly more elastic than a lower molecular weight EAK16-II of the same structure. Although both researches indicate the influence of the molecular sizes of the peptides, the relation between molecular size and self-assembling behavior still remains unclear. Further study is needed to clarify the definite relationship. The effect of the variation of hydrophobic side chains on self-assembly was also investigated based on the KFE series. The three sequences of KFE12, KIE12, and KVE12 were studied first and demonstrated the importance of hydrophobicity for 3D-gel matrix formation [28]. In addition to the hydrophobicity, the sequence and steric size of the hydrophobic residues were also found to cause variations in peptide conformation and create significant differences in the matrix morphology of self-assembled peptides [30]. Hydrogel 3D-matrices from β-sheet peptide-00 (this peptide corresponds to KFE16-II. Amino acid sequence; see Table 1) had many terminations, fine mesh size, and uniform diameter of ~ 20 nm. On the other hand, 3D-matrices formed by peptides-81 and -82a had fewer terminations or branch points, a large mesh size, and thicker diameter than that from peptide-00. Peptide-83 did not form hydrogel, and existed in random coil structures. The authors explained these differences in self-assembly of β-strands as the different steric topologies of non-polar side chains as follows. Similar steric sizes of non-polar residues for peptide-00 permit strong hydrophobic interactions anywhere along the peptide, although steric buldges in peptides-81 and -82a create a much more selective environment for hydrophobic interactions to take place. Moreover, there is a very high entropic barrier for peptide-83 β-strands to bind (Fig. 6). Other peptides with hydrophobic and hydrophilic patterns have also been designed to learn about the relationship between peptide primary structure and the consequent morphology of the self-assembled nanostructures, e.g., from the view point of the effects of turn sequence (MAX1 and MAX1(L)) [31], and terminal Pro residues (FF, FI, VI) [32].

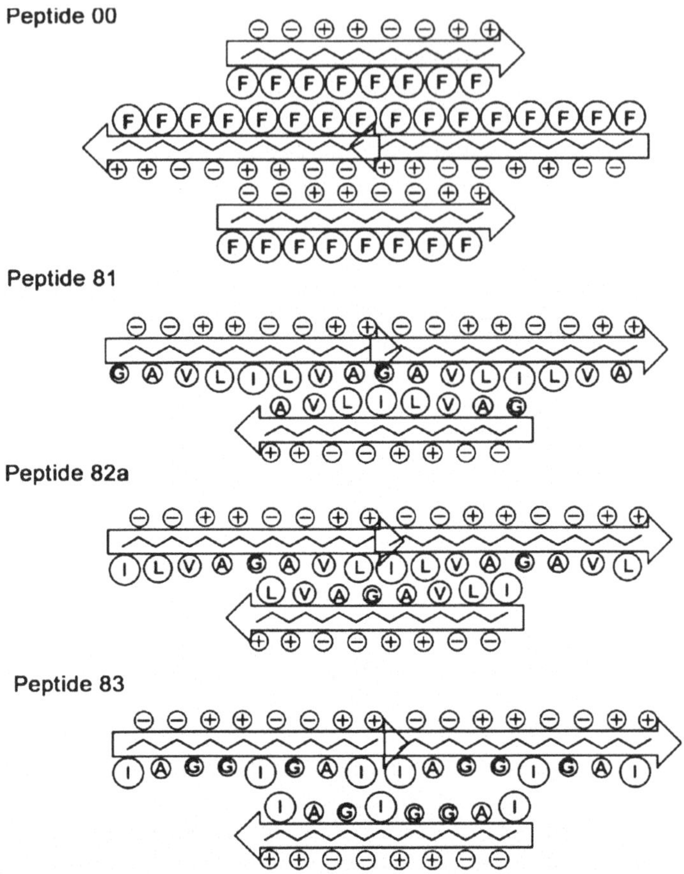

Fig. 6 Schematic representation of the packing between β-sheets of peptides with various hydrophobic bulges. (Adapted from [30])

The second type of self-assembling building unit is "peptide-amphiphile (PA)", which consists of a hydrophobic alkyl tail coupled to a peptide block and includes β-sheet forming segments and charged residues (type II in Fig. 1). These PA molecules have been developed by Stupp et al. [33–38]. The tendency of the alkyl tail to aggregate in aqueous solution is the driving force for self-assembly. It has been found for many PAs that the self-assembly leads to the formation of well-defined nanofibers in excess of 1 µm long and 5–10 nm in diameter upon application of a trigger, such as a change in pH or ion-concentration. In this nanofiber, the alkyl chains are in the core while the hydrophilic peptide region is displayed on the nanofiber surface, and the β-sheet formation between peptide region stabilizes the nanofiber structure. Hartgerink et al. investigated the role of hydrogen bonding in self-assembly of PAs in detail by using a series of 26 PA derivatives, including

Table 2 Sequences of *N*-methylated and alanine-substituted PAs (**1–26**), consist of a C16 hydrophobic tail, a glycine containing region, and a charged head group, designed by Hartgerink et al. (Adapted from [36])

Position of G residues 1 2 3 4 5 6 7

	G Position						
	1	2	3	4	5	6	7
PA1	G	G	G	G	G	G	G
PA2	G	G	G	G	G	G	^{NMe}G
PA3	G	G	G	G	G	^{NMe}G	^{NMe}G
PA4	G	G	G	G	^{NMe}G	^{NMe}G	^{NMe}G
PA5	G	G	G	^{NMe}G	^{NMe}G	^{NMe}G	^{NMe}G
PA6	G	G	^{NMe}G	^{NMe}G	^{NMe}G	^{NMe}G	^{NMe}G
PA7	G	^{NMe}G	^{NMe}G	^{NMe}G	^{NMe}G	^{NMe}G	^{NMe}G
PA8	^{NMe}G	^{NMe}G	^{NMe}G	^{NMe}G	^{NMe}G	^{NMe}G	^{NMe}G
PA9	^{NMe}G	G	G	G	G	G	G
PA10	G	^{NMe}G	G	G	G	G	G
PA11	G	G	^{NMe}G	G	G	G	G
PA12	G	G	G	^{NMe}G	G	G	G
PA13	G	G	G	G	^{NMe}G	G	G
PA14	G	G	G	G	G	^{NMe}G	G
PA15	^{NMe}G	^{NMe}G	G	G	G	G	G
PA16	^{NMe}G	^{NMe}G	^{NMe}G	G	G	G	G
PA17	^{NMe}G	^{NMe}G	^{NMe}G	^{NMe}G	G	G	G
PA18	^{NMe}G	^{NMe}G	^{NMe}G	^{NMe}G	^{NMe}G	G	G
PA19	^{NMe}G	^{NMe}G	^{NMe}G	^{NMe}G	^{NMe}G	^{NMe}G	G
PA19	^{NMe}G	^{NMe}G	^{NMe}G	^{NMe}G	^{NMe}G	G	G
PA20	G	G	G	G	G	G	A
PA21	G	G	G	G	G	A	G
PA22	G	G	G	G	A	G	G
PA23	G	G	G	A	G	G	G
PA24	G	G	A	G	G	G	G
PA25	G	A	G	G	G	G	G
PA26	A	G	G	G	G	G	G

19 *N*-methylated variants and 7 alanine mutants, as shown in Table 2 [36]. As a result, they found that the four amino acids closest to the alkyl core of the nanofiber form *β*-sheet hydrogen bonds, and that disruption of these hydrogen bonds causes the PA to self-assemble into spherical micelles instead of nanofibers. Since the various functional peptides, such as bioactive epitopes, can be easily incorporated into the PA self-assembling system, these PA molecules are finding many applications in the fields of nanotechnology and biotechnology (Sects. 2.4 and 2.5). It has also been shown that two different PA molecules with complementary charges can be co-assembled into the same nanofiber [37, 38].

The third type of self-assembling units are peptidomimetics [39–42], in which the amphiphilic oligopeptides are combined with hydrophobic synthetic compounds (Type III in Fig. 1). For example, Kelly et al. prepared a dibenzofuran-based simple peptidomimetic and showed that the peptidomimetic formed polymorphic *β*-sheet assemblies including protofilaments, filaments, ribbons, and fibrils [39]. On the basis of the TEM, AFM, and X-ray analyses, they proposed the hierarchical self-assembly of the peptidomimetic as follows. At first, the peptidomimetic self-assembles affording quaternal structures with an average width of 25 Å and height of 6 Å, corresponding to the width and height of a single cross-β monolayer. Face-to-face dimerization of the cross-β monolayer mediated by hydrophobic interactions provides a 10–12 Å thick bilayer, referred to as protofilaments. These protofilaments further elongate and interact via lateral association to form filaments with a width of 50–60 Å, composed of two protofilaments. Finally, the filaments can associate to form ribbon and fibrils (Fig. 7). In this molecular system, the dibenzofuran-template not only pre-organizes the peptidomimetic for dimerization and further assembly, but also provides a hydrophobic edge to promote protofilament and filament assembly. Mihara et al. synthesized the peptide (Ad-2α), which composed of two amphiphilic α-helix segments, each modified with a 1-adamantane carbonyl group (Ad) at the N-terminus as a hydrophobic defect [40]. This peptide folded into a 2α-helix structure at initial state, but gradually self-assembled into amyloid-like fibrils (ca. 10 nm width) with α-to-β structural transition in neutral aqueous solution (e.g., Fig. 11). In this case, exposed Ad hydrophobic defects in the peptide were responsible for both the structural transition and self-assembly. They have also developed the Ad-2α self-assembling system to enable the co-assembly from three- or four-different peptides into the same nanofiber by complementary electrostatic interactions, which seems to be important step for the design of multi-functional nanofiber materials [41].

These studies will undoubtedly make an important contribution not only to our understanding of the mechanism of the peptide aggregation as a simple

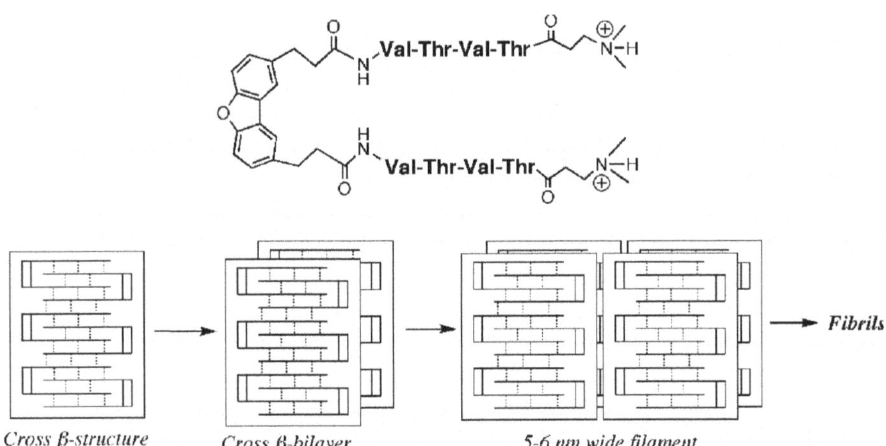

Fig. 7 Schematic illustration of hierarchical self-assembly envisioned for 2,8-dibenzofuran-based peptidomimetic in water. (Adapted from [39])

and useful model, but also in establishing a novel molecular self-assembling system for the fabrication of the designed 3D-nanoarchitectures.

2.2
Stimuli-Responsive Self-Assembling System into Nanofibers

In Sect. 2.1 above, we focused on the formation and regulation of β-sheet-based nanostructures, including nanofibers, from chemically synthesized artificial peptides. Generally, β-strand peptides self-assemble uncontrollably, affording β-sheet-rich materials that may be thermodynamically stable and display limited responsiveness. Therefore, it is also important to establish facile and practical guidelines to control the spontaneous peptide self-assembling process in addition to nano-structural control, that is temporal control, in order to advance a bottom-up nanobiotechnology (programmable peptide self-assembly). Some groups have reported the construction of stimuli-responsive peptide self-assembling systems [43–45].

Imperiali's group reported the photo-controlled peptide self-assembling system into fibrillar structures [44]. They designed amyloidogenic peptide (PrP174-195 fragment) modified with a "fibril-inhibitory unit (*N*,*N*-dimethyl-ethylendiamine-tagged oligo(Asn) (DMDA-Asn))" through a photolabile linker (3-amino-3-(2-nitrophenyl)-propionic acid (ANP)) (Fig. 8). HPLC and TEM analyses clearly showed that the DMDA unit stabilized the peptide as a soluble species without UV-irradiation for at least 12 h

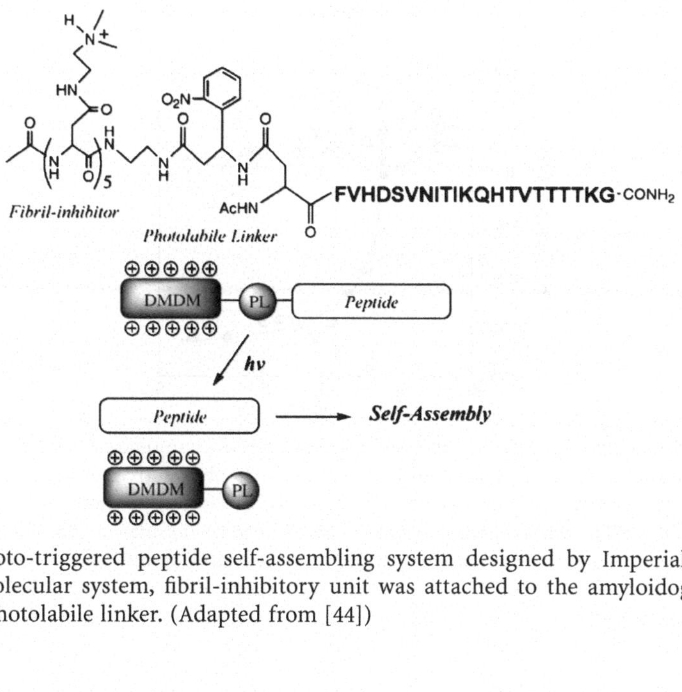

Fig. 8 Photo-triggered peptide self-assembling system designed by Imperiali's group. In this molecular system, fibril-inhibitory unit was attached to the amyloidogenic unit through photolabile linker. (Adapted from [44])

in a phosphate buffer at pH 7.5. However, when the solubility DMDA unit was photo-dissocinated from the amyloidogenic peptide, the photo-product showed rapid aggregation into fibrillar nanostructure. Thus, photo-triggered self-assembly was successfully constructed by controlling the amphiphilicity of the target peptide. Controlling the amphiphilic balance of the peptide by external stimuli, such as light, is one of key technology to accomplish a temporal control of peptide self-assembly.

We have also fabricated a unique and programmable peptide self-assembly system by using poly(ethylene glycol)(*PEG*)-attached amphiphilic oligopeptide, which shows rapid self-assembly into well-organized β-sheet nanofibers in response to an enzymatic reaction [45]. This peptide (LKL16-*PEG*) was composed of β-sheet forming amphiphilic peptide (LKL16), thrombin-cleavable amino acid sequence (VPRGS), and hydrophilic *PEG* segment (Fig. 9A). MALDI-TOFMS spectra clearly showed that the site-specific di-

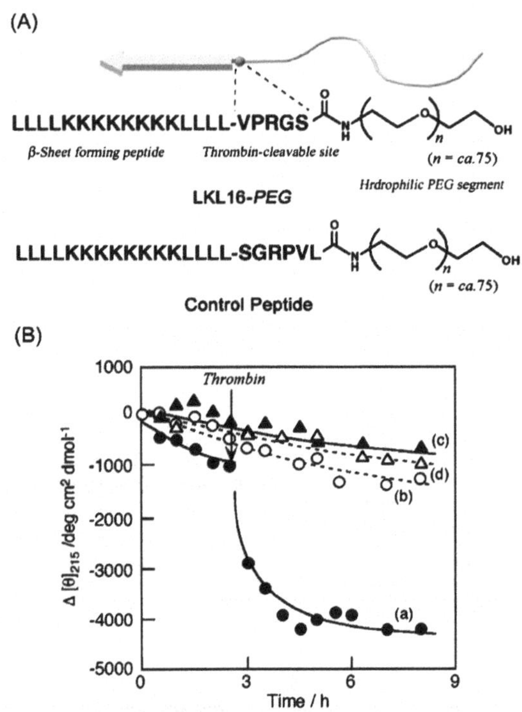

Fig. 9 **A** Molecular structures of poly(ethylene glycol)-conjugated β-sheet peptides, LKL16-*PEG* and its control peptide with scrambled thrombin-active sequence (SGRPVL), which were used as building blocks for self-assembly. **B** Time dependences of $\Delta[\theta]_{215}$ for LKL16-*PEG* (a and b) and control peptide (c and d) with (*solid lines*; a and c) and without the addition of thrombin (*dashed lines*; b and d) under above condition. The *arrow* indicates the time at which thrombin was added in the peptide-solution. [peptide] = 40 μM. (Adapted from [45])

gestion of LKL16-*PEG* took place smoothly at the R–G bond within 30 min and that, as a result, the LKL16-VPR peptide was generated by detaching the *PEG*-based unit. The conformational properties of the LKL16-*PEG* in response to the enzymatic reaction were subsequently investigated by means of CD spectroscopy in buffer at pH 9.0. With freshly prepared sample solution, the CD spectrum gave mixed patterns of α-helix and random coil structures. This α-helical conformation of LKL16-*PEG* was maintained even after 8 h-incubation. On the other hand, interestingly, the addition of thrombin was found to cause a remarkable α-to-β structural transition (Fig. 9B). Together with such conformational change by thrombin addition, AFM measurements demonstrated the formation of well-organized nanofibers with nearly uniform diameters of ca. 5–6 nm, although the LKL16-*PEG* without thrombin-digestion formed amorphous aggregates (Fig. 10). Thus, the observed enzyme-triggered self-assembly of LKL16-*PEG* into a β-sheet nanofiber would be driven by the detaching of the GS-*PEG* unit from the original peptide, permitting attractive hydrophobic and van der Waals forces among the resultant LKL16-VPR peptides to dominate. More recently, we have successfully accomplished OFF-to-ON switching of peptide self-assembly into β-sheet nanofibers by thrombin digestion, at which a D,L-alternating oligolysine ($(K_DK_L)_3$) unit was employed as a hydrophilic and non-conformational tail instead of a *PEG*-based hydrophilic tail [46]. Mutter's group used enzymes to trigger an acyl migration that allowed switching of a conformation and self-assembly of various peptides [47].

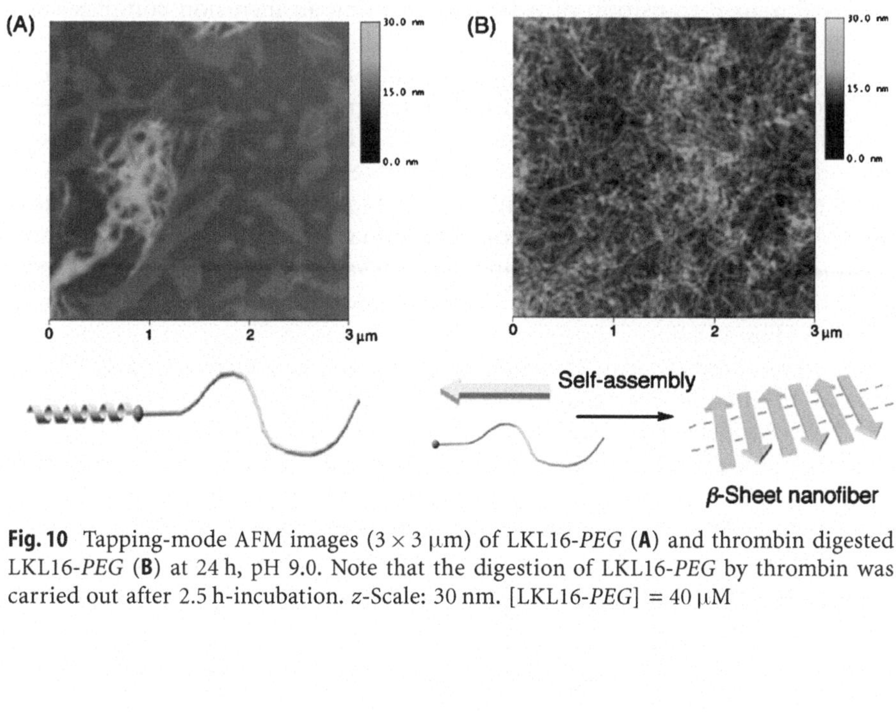

Fig. 10 Tapping-mode AFM images (3 × 3 µm) of LKL16-*PEG* (**A**) and thrombin digested LKL16-*PEG* (**B**) at 24 h, pH 9.0. Note that the digestion of LKL16-*PEG* by thrombin was carried out after 2.5 h-incubation. *z*-Scale: 30 nm. [LKL16-*PEG*] = 40 µM

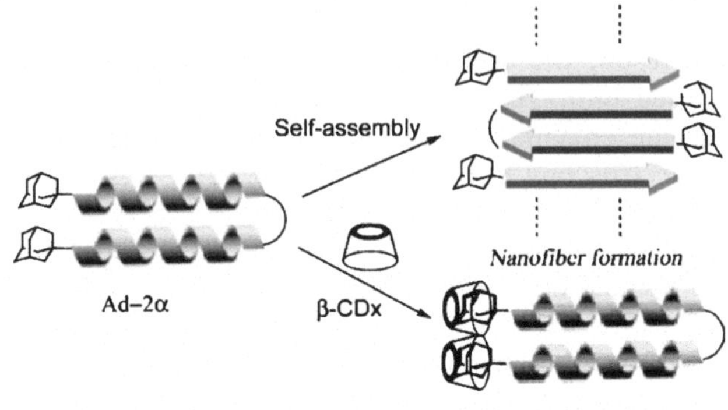

Fig. 11 Schematic illustration of the β-cyclodextrin(β-CDx)-controlled conformation and self-assembly of Ad-2α peptide established by Mihara et al. Self-assembly of Ad-2α with hydrophobic defects into β-sheet nanofibers was prevented by the complexation with β-CDx. (Adapted from [40])

Guest additive-regulated peptide self-assembling systems have been reported [40, 48–50]. For example, Mihara's group fabricated the β-cyclodextrin (β-CDx)-controlled peptide self-assembling system by using above mentioned Ad-2α having 1-adamantane carbonyl groups [40]. This peptide gradually self-assembles into amyloid-like fibrils with α-to-β structural transition due to strong hydrophobicity of terminal Ad groups in neutral aqueous solution. Interestingly, the addition of β-CDx (5.0 equiv to the Ad group) inhibited α-to-β transition of Ad-2α by forming an inclusion complex specific to the Ad group, which diminishes or shields the hydrophobic defects in Ad-2α (Fig. 11). In contrast, guest (metal)-induced nanofiber formation was archived by the same group using a designed β-sheet peptide with His residues at the both N- and C-termini, HPKFKIIEFEPH [48]. The peptide was in random coil in a neutral buffer, but underwent the structural transition to a β-sheet structure and self-assembled into tape-like nanofibers in response to Cu^{2+} ion. In this case, interaction of terminal His residues with Cu^{2+} ions acted in an important role in causing such self-assembly, which was also supported by the fact that the tape-like nanofibers were broken by the addition of EDTA (Fig. 12).

Photo-isomerization of the exogenous anionic azobenzene ($Az(COO^-)_2$) has also been employed to control a self-assembling process of cationic peptide [49]. Azobenzene is well known to change the overall geometry from planar to non-planar and the polarity of azo-linkage by *trans*-*cis* photoisomerization. This polarity change of azobenzene can be transferred to the target peptide by complexation as a slight change of total hydrophobicity. Figure 13 shows the transition kinetics of the cationic LKL16 peptide into the β-sheet structure at pH 9.2 with *trans*-isomer and *cis*-isomer. In the presence

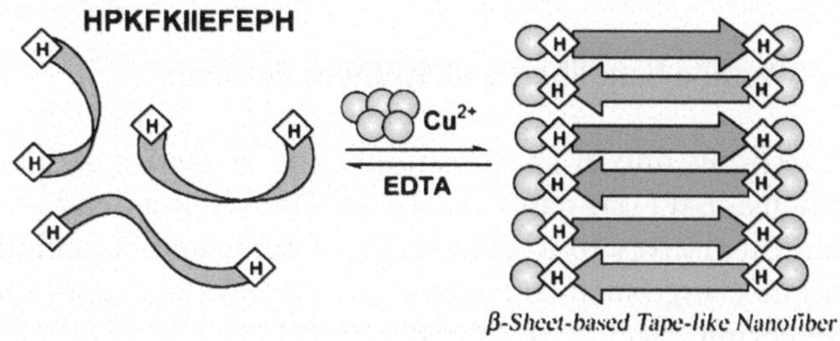

Fig. 12 Schematic illustration of the tape-like nanofiber formation of HPKFKIIEFEPH peptide induced by Cu^{2+} and the destruction by EDTA. (Adapted from [48])

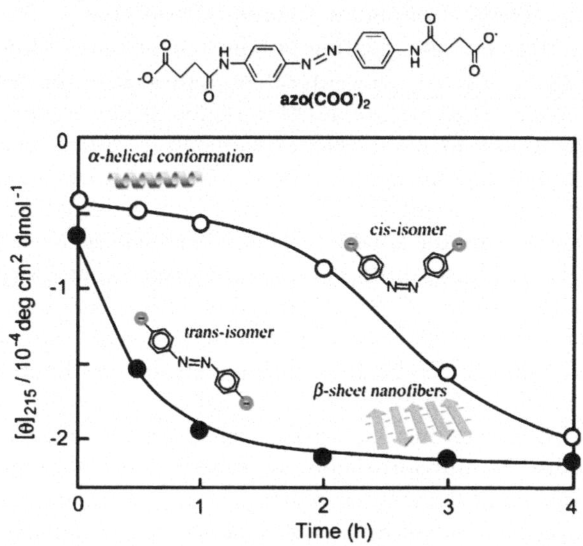

Fig. 13 Time dependences of $[\theta]_{215}$ values in CD spectra of LKL16 in the presence of *trans*-azo$(COO^-)_2$ (•) and *cis*-azo$(COO^-)_2$ (◦) in borate buffer (containing 5% TFE) at pH 9.2. The peptides were incubated for the time indicated (0–4 h). [LKL16] = 40 μM. [azo$(COO^-)_2$] = 80 μM. (Adapted from [49])

of *cis*-isomer, the $[\theta]_{215}$ value in CD spectra decreased sigmoidally and the β-sheet formation was accelerated at around 2–3 h. In contrast, a remarkable acceleration of the α(random)-to-β transition was observed in the presence of *trans*-Az$(COO^-)_2$. It seems that the hydrophobic *trans*-Az$(COO^-)_2$ (0–0.5 D) triggered a rapid formation of the β-sheet nucleus, which leads to the disappearance of the lag-phase before conformational transition, by increasing the total hydrophobicity of LKL16, in contrast to the case of relatively hydrophilic *cis*-isomer (3 D).

The ability to regulate the peptide self-assembly temporally should be useful in establishing a programmable molecular assembling system for the fabrication of designed 3D-nanoarchitectures at controlled timing.

2.3 Hybridization of Peptide Nanofibers and Synthetic Polymers

Peptide self-assembly into β-sheet nanofiber is attractive strategy to construct one-dimensional (1D) nanostructures. These self-assembled 1D-nanostructures lend themselves to a wide range of applications, including use in biomedical engineering (such as tissue engineering) (Sect. 2.4) and in nanotemplate engineering (Sect. 2.5). Transferring these organization principles towards the structuring of synthetic polymers at the nano-scale would be innovative, and expand the potential applications drastically of self-assembled nano-architectures. Thus, peptide-polymer hybrids will open a new generation of self-assembled nanofibers with unique properties.

In order to fabricate the peptide-polymer hybrid nanofibers, two approaches have been devised. One attractive approach is the work on a covalent attachment of synthetic polymers to oligopeptide-organizer units, i.e., nanofiber- or nanotube-formable short oligopeptides. Burkoth et al. utilized a conjugate of poly(ethylene glycol) with the C-terminus of Aβ(10–35) fragment of the amyloid β sequence (Fig. 14A) [51]. The Aβ(10–35)-*PEG* self-assembled into water-soluble amyloid-like uniform aggregates with diameters of > 8 nm at pH 5.7. Messersmith et al. demonstrated that shorter oligopeptides (7-mers and 11-mers; *EG12*-Q7 and Q11-*PEG*) also have the inherent potential to self-organize *PEG* or oligo(ethylene glycol) into water-soluble nanofibers (Fig. 14B) [52]. In both cases, interestingly, the lateral fiber aggregations were obviously suppressed, and the nanofibers existed without pairing under TEM observations. Generally, self-assembled β-sheet nanofibers have a propensity for uncontrollably aggregating into disordered tangles. Thus, the PEGylation of β-sheet nanofiber-forming peptides enable the modulation of nano-scale association of the nanofibers, particularly their lateral packing. It should be noted that the PEGylation of β-sheet peptide sometimes changes its primary self-assembling property as described in Sect. 2.2, which is probably based on the difference in aggregation tendency of the peptide-organizer unit.

In order to strengthen a self-assembling tendency of peptide-organizer unit in peptide-*PEG* hybrid, Börner et al. employed the carbazole-template pre-organized oligopeptide instead of linear oligopeptide (Fig. 15) [53]. It has been shown by Kelly et al. that the restriction of the conformational freedom of oligopeptides and their pre-organization into optimized geometry strongly promotes the anti-parallel β-sheet formation [39]. CD, AFM, and TEM measurements revealed that the *PEG-block*-(template-oligopeptide) formed uniform β-sheet nanofibers with a height of ca. 1.4 nm and a maximum length in the range of a micrometer under neutral conditions (Fig. 15). By comparison of the pre-organized conjugate with a linear *PEG-block*-oligopeptide analog, it was demonstrated that pre-organization strongly enhanced the aggregation behavior. The author noted that the tendency of such oligopeptide

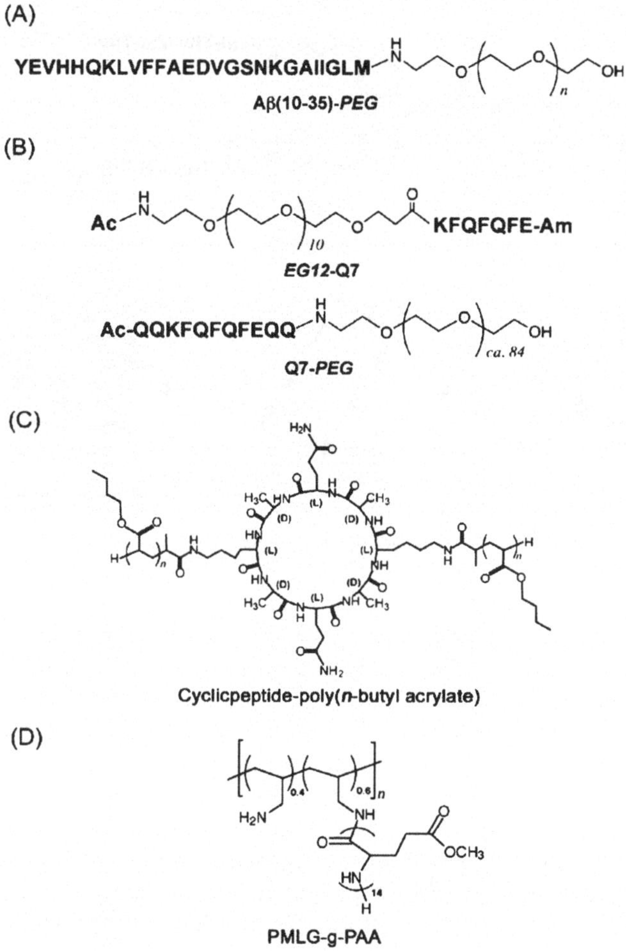

Fig. 14 Chemical structures of various peptide-polymer hybrids that self-assemble into β-sheet-based fibrous nanostructures

organizer units to form well-defined nanostructures can be potentially modulated over a wide range of tailor-made aggregators with tunable aggregation strength and responsiveness as well as different target nanostructures. Based on this concept, different types of peptide-polymer hybrid have already been prepared by the same research group [54]. The peptide-polymer hybrid was composed of a (D-alt-L)-cyclicpeptide as aggregator domain and a poly(*n*-butyl acrylate)-graft chain (Fig. 14C). The controlled self-assembly of the hybrid led to uniform nano-scale tube structure in organic solvent, in which a tubular peptide-core was built by stacking the cyclic peptides and wrapping poly(*n*-butyl acrylate)-graft chains around the peptide tube. In contrast, we prepared an artificial peptide-grafted synthetic polymer: poly(γ-methyl-

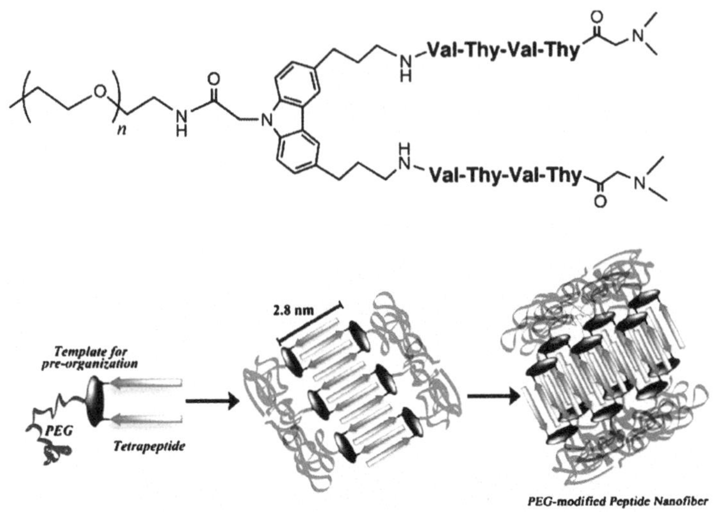

Fig. 15 Proposed structure for the aggregation of *PEG-block*-(template-oligopeptide) into fiber-like aggregates showing a peptide core and a *PEG* shell. (Adapted from [53])

L-glutamate)-grafted polyallylamine (PMLG-*g*-PAA) (Fig. 14D) [55, 56]. The PMLG segment with relatively short chain length ($n = 14$), which was employed as a secondary structure module, provides the hydrophobic driving force for self-assembly in water. The secondary and quaternary structures of the PMLG-*g*-PAA could be easily controlled by manipulating the pH, solution composition, and salt concentration, reflecting the propertiy of synthetic PAA polymer unit. Only under the condition of low pH (< 8) or that of low salt concentration (< 30 mM) did the electrostatic repulsive force based on the protonated amino groups of PAA permit the controlled growth of PMLG-*g*-PAA-aggregates and the tight packing of hydrophobic PMLG chains, resulting in induction of highly-ordered nanofiber formation with α-to-β structural transition (Fig. 16). On the other hand, at a higher pH (> 8) region or higher salt concentration (> 30 mM), the PMLG-*g*-PAA was assembled rapidly with α-helical conformation and formed poorly organized amorphous aggregates. At high pH or high salt concentration, the positive charges present on the PMLG-*g*-PAA will be screened or disappeared (pK_a of amino groups of PAA segment: ca. 8.0). This increased tendency for intermolecular interactions between PMLG-*g*-PAAs, which cause rapid aggregation of PMLG-*g*-PAA, is due to a drastic decrease in electrostatic repulsion and an increase in hydrophobicity. Thus, this result demonstrated the possibility of a peptide-polymer hybrid to regulate the self-assembling behaviour of artificial peptides depending on the property of synthetic polymer segments.

Another approach for fabricating the peptide-polymer hybrid nanofibers is the polymerization method at the surface of self-assembled β-sheet nanofibers (or nanotubes). In the β-sheet structure, adjacent β-strands align

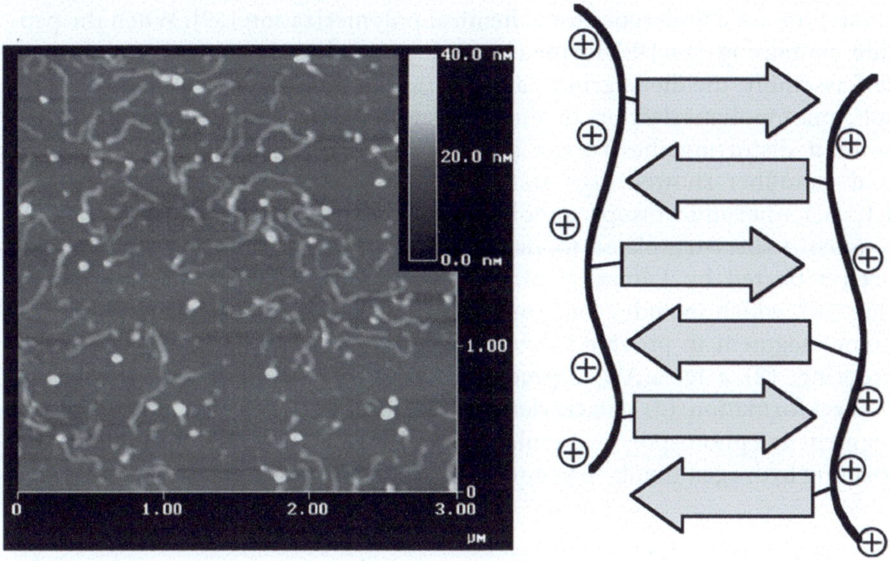

Fig. 16 Tapping mode AFM image ($3 \times 3\ \mu m$) for PMLG-*g*-PAA after incubation for 48 h in water/TFE (8 : 2 v/v) (pH 4.0) at room temperature. The PMLG-*g*-PAA formed nanofiber structures with conformational transition from α-helix to antiparallel β-sheet structure. *z*-Scale: 0–40 nm. [MLG] = 1.1×10^{-4} M

at a distance of 4.3–4.8 Å depending on the amino acid sequence, typically 4.8 Å [57]. Mori et al. incorporated a diene as the polymerizable group into the side chain of Lys residue (Lys(DE)), which is located in the center of the β-sheet peptide sequence (KFE type) in order to cross-link β-strands (Fig. 17) [58]. It is known that diene aligned at a distance of 4.9–5.1 Å in the

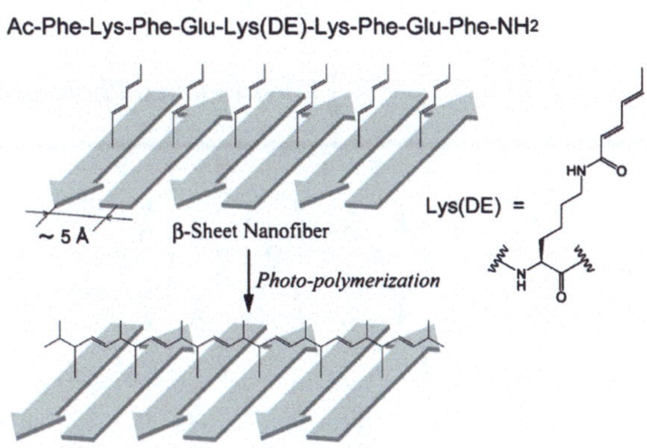

Fig. 17 Radical polymerization of diene (DE) groups attached to the β-sheet peptides at the surface of self-assembled nanofibers. (Adapted from [58])

crystalline state undergoes topochemical polymerization [59]. When the peptide containing Lys(DE) formed an anti-parallel β-sheet structure through self-assembly, the diene groups aligned ca. 5 Å apart on the β-sheet. As a result, photopolymerization of the diene groups was successfully proceeded without distorting the β-sheet structure (Fig. 17). The obtained polymerized nanofiber showed high stability against acid-denaturation and added 1,1,1,3,3,3-hexafluoroisopropanol (HFIP).

Such β-sheet-templated topochemical polymerization of diacetylenes has been reported by Jahnke et al. [60]. They designed the macromonomer (Fig. 18), which includes four functional units; (1) a hydrogenated poly(isoprene) segment to provide solubility in organic solvents and prevent global ordering; (2) a tetra(Ala) segment to induce anisotropic self-assembly by β-sheet formation; (3) a diacetylene moiety integrated directly into the peptide segment as a photo-polymerizable unit; and (4) an NHAc end group capable of forming hydrogen bonds to promote a parallel arrangement of the molecules.

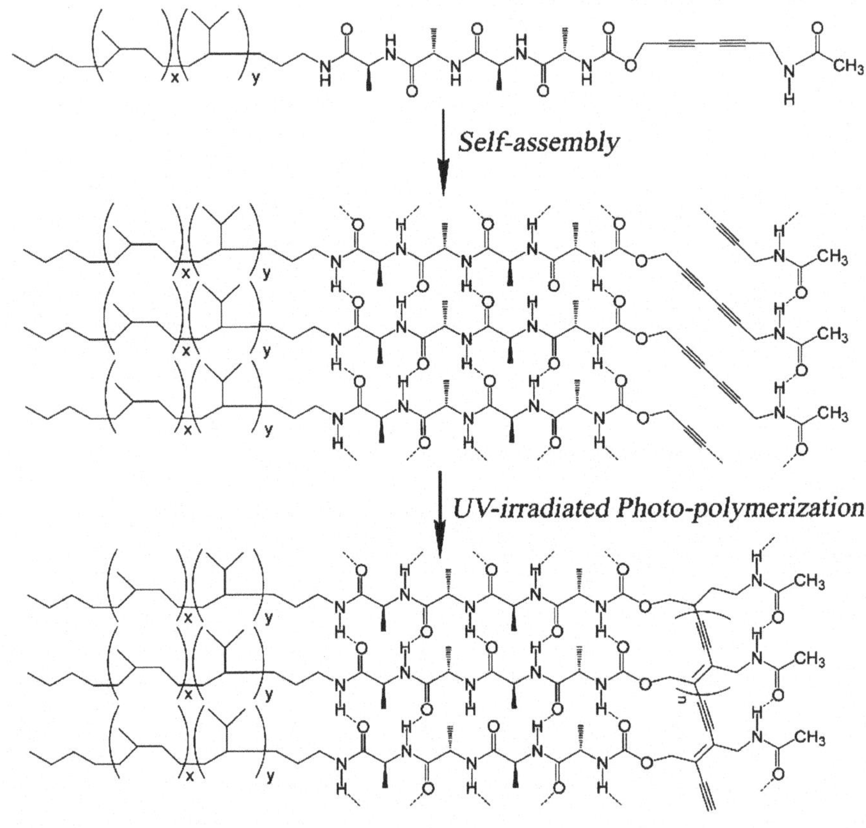

Fig. 18 Self-assembly and topochemical polymerization of diacetylenes functionalized with a β-sheet-forming oligopeptide. (Adapted from [60])

The macromonomer self-assembled into a supra-molecular fibrillar structure with a double-helical topology that was then converted into the corresponding poly(diacetylene) by UV irradiation (Fig. 18). Such highly-conjugated organic-nanofiber is expected to exhibit electronic and optical properties of interest for applications ranging from molecular-wires to biosensors based on extended π-electron delocalization along their backbones.

Atom transfer radical polymerization (ATRP) is also an interesting strategy to construct peptide-polymer hybrid 1D-arrays. Graft-polymerization of *N*-isopropylacrylamide was carried out from the surface of self-assembled peptide nanotube in water, and as a result, cyclicpeptide-poly(*N*-isopropylacrylamide) hybrid nanotube was successfully prepared (Fig. 19) [61]. In other words, this research indicates that surface chemistry of 1D-peptide nano-assembly can be adjusted by the grafted synthetic polymers, such as thermo-sensitive PNIPAM and polyelectrolytes.

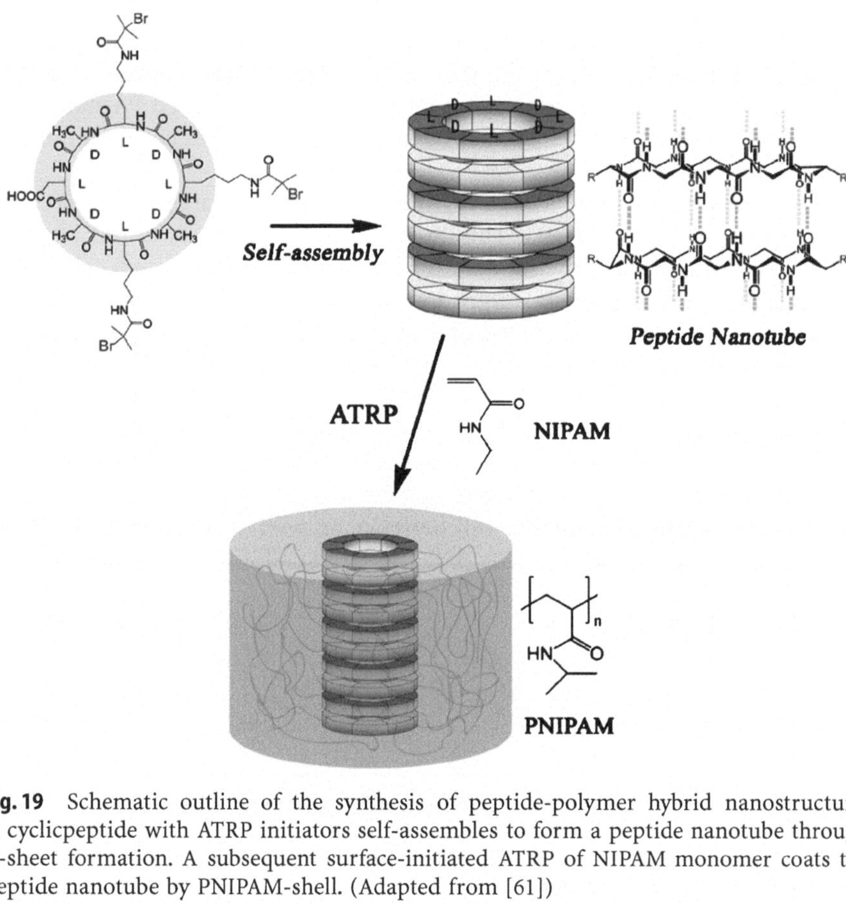

Fig. 19 Schematic outline of the synthesis of peptide-polymer hybrid nanostructure. A cyclicpeptide with ATRP initiators self-assembles to form a peptide nanotube through β-sheet formation. A subsequent surface-initiated ATRP of NIPAM monomer coats the peptide nanotube by PNIPAM-shell. (Adapted from [61])

2.4 Applications in Biomedical Engineerings

Self-assembling β-sheet peptides have provided the de novo designed nano-scaffolds with potential applications in biomedical engineering, such as tissue engineering, biomineralization, and ultimately, regenerative medicine [33, 35, 62–71]. For example, Stupp et al. used a cylindrical β-sheet of nanofibers prepared through self-assembly of PA molecules to serve as a template for the mineralization of hydroxyapatite (HA) crystals in the correct crystallographic orientation [33]. As we introduced in Sect. 2.1, these PA molecules consist of a hydrophobic alkyl tail, covalently coupled to a charged peptide head group containing a β-sheet forming segment, and a bioactive segment. By incorporating a phosphorylated Ser within the peptide region, calcium or other ions can be attached, which enables direct mineralization of HA (Fig. 20). At the lowest structural level in bone tissue is the organization of collagen fibrils with respect to HA crystals. The collagen fibrils are formed by self-assembly of collage triple helices and the HA crystals grow inside grooves and channels of these fibrils in such a way that their *c*-axes are oriented along the long axes of the fibrils [64, 65]. In the case of PAs, preferential alignment of the HA crystallographic *c*-axis with the long axis of self-assembled PA-nanofiber with ca. 8 nm in diameter was observed as well as the case of collagen fibril in bone [33].

The C-terminus of the PA was further modified with a three amino acid moiety of Arg-Gly-Asp (RGD). This RGD sequence has been found by Pierschbacher et al. to play an important role in integrin-mediated cell adhesion [66]. Since the peptide head group is displayed at high density along the nanofiber periphery in this molecular system, incorporation of RGD motif into the PA system promotes cell attachment. In fact, the efficient adhesions of mesenchymal stem cells [67] and bladder smooth muscle cells [68] to the RGD-containing PA-nanofibers have been demonstrated. Bioactive sequence of IKVAV, which is found in laminin and known to promote neurite sprouting and to direct neurite growth, has also been employed in these PA systems to act as epitope for influencing cell behavior as well as RGD sequence. The

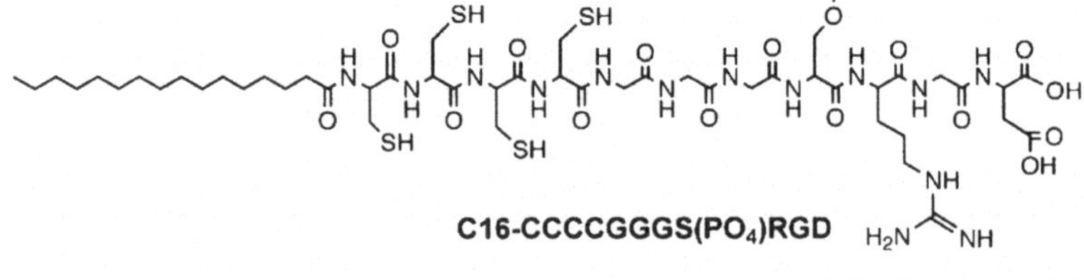

Fig. 20 Chemical structure of self-assembling peptide-amphiphile (PA) with bioactive sequence (designed by Stupp et al.)

IKVAV-containing PA-nanofiber scaffold induced very rapid differentiation of neural progenitor cells into neurons based on the amplification of bioactive epitope presentation to cells by the nanofibers (roughly estimated to be 7.1×10^{14} IKVAV epitopes/cm^2) [35].

Another kind of peptides containing 12–20 alternating hydrophobic and hydrophilic amino acids are induced to self-assemble into nanofibers, and ultimately form self-supported hydrogels with great potential applications in tissue engineering. For instance, Kisiday et al. proved the usefulness of peptide hydrogel scaffold from 12-mers amphiphilic peptide, KLD-12 (AcN-KLDLKLDLKLDL-CNH$_2$), for cartilage repair [69]. During 4 weeks of culture in vitro, chodrocytes seeded within the hydrogel, retained their morphology, and developed a cartilage-like extracellular matrix rich in proteoglycans and type II collagen. Such β-sheet-3D-scaffold made from RADA16 [20] has been used for controlled differentiation of putative adult rat progenitor cells [70]. Another 20-mer amphiphilic peptide, consisting of alternating K and V residues and flanking a type II′ β-turn (MAX1 in Table 1), has been reported to form rigid hydrogel in Dulbecco's modified eagle's medium (DMEM). This hydrogel fostered the ability of the NIH3T3 cells to attach to the hydrogel scaffold in the absence and presence of serum proteins, and was able to support fibroblast proliferation to confluency [71].

The cytocompatibility and mechanical strength of these peptide 3D-scaffolds, as well as the easy incorporation of bioactive motifs to the scaffolds, make them attractive candidates for use in biomedical engineering.

2.5 Applications for Nano-Templates

Self-assembled peptide nano-objects, including nanofibers and nanotubes, are expected to possess potential as templates for producing nano-scale electronic components, and accomplishing dimension-regulated nano-organization of functional molecules such as metal particles, proteins, and synthetic polymers.

Gazit et al. showed the synthesis of metal nanowire using peptide nanotube, which was prepared through self-assembly of very short H$_2$N-Phe-Phe-COOH dipeptide [72]. This dipeptide appeared to be highly soluble in the organic solvent of HFIP ($\geq$ 100 mg/ml), but showed a rapid assembly into discrete and stiff nanotubes with a diameter of ca. 100–150 nm and a β-sheet-like conformation after dilution into aqueous solution at a final concentration of μM range. Reduction of ionic silver (Ag^+) within the nanotubes, followed by enzyme degradation of the peptide mold, enabled the production of discrete silver nanowire with ca. 20 nm diameter (Fig. 21).

Peptide nanofibers were also utilized as nano-template for the synthesis of metal nanowires and for the spatial control of inorganic nanoparticles [73, 74]. Wang et al. prepared β-sheet-based peptide nanofiber with

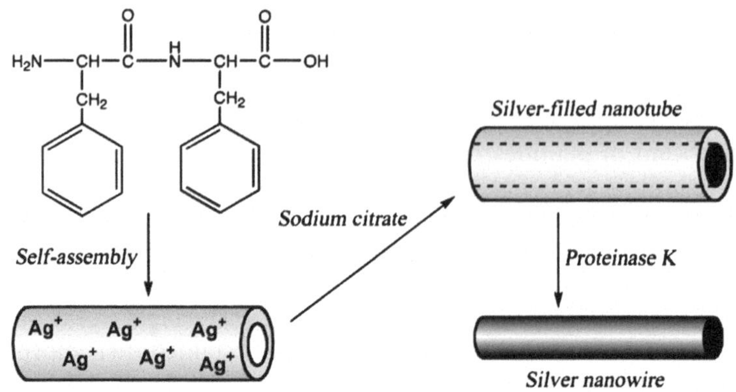

Fig. 21 Casting of silver nanowires with the peptide nanotubes. The nanowires were formed by the reduction of silver ions within the nanotubes, followed by enzymatic degradation of the peptide mold. (Adapted from [72])

positive charge from 12-mers synthetic peptide (T1: H_2N-RGYPWAGDYNYP-COOH). It is well known that metal colloidal nanoparticles are usually negatively charged due to the adsorption of anionic stabilizer. Therefore, the metal nanoparticles would self-organize at the nanofiber surfaces through electrostatic interactions. In fact, the author demonstrated one-dimensional arrays of Au and Pd nanoparticles along the nanofiber-template from TEM measurements. In addition, when a $PdCl_2$ solution was repeatedly added to a suspension of T1-Pd hybrid nanofiber followed by reduction of Pd^{II} ions with hydrogen ($Pd^{II} + H_2 \rightarrow Pd^0 + 2H^+$), Pd nanoparticles grew and more Pd particles aggregated on the template, resulting in the formation of Pd nanowires [74]. These strategies are limited, however, to water-soluble nanoparticles, be-

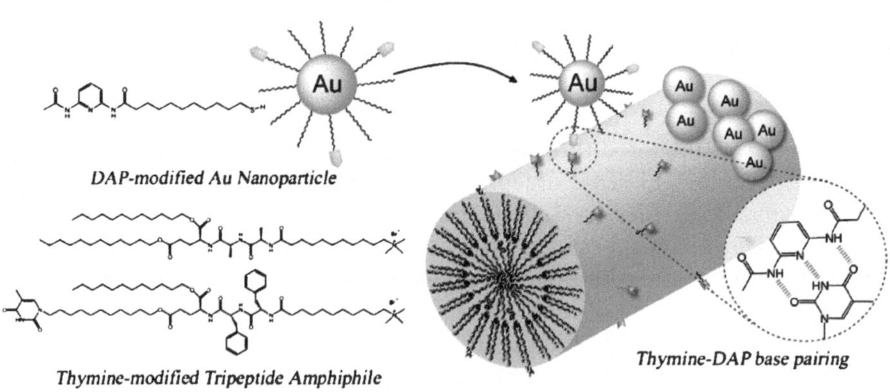

Fig. 22 Schematic illustration of the immobilization of diaminopyridin-functionalized Au nanoparticles onto thymine-modified self-assembled β-sheet nanofiber through complemental base-pairing in apolar solvent. (Adapted from [75])

cause the templates used require water as a solvent. To address this issue, Stupp et al. devised a strategy for the creation of one-dimensional assemblies of lipophilic inorganic nanoparticles in apolar solvents by using peptide-based nanofibers with surface-binding motifs [75], as illustrated in Fig. 22. Lipophilic Au nanoparticles modified with diaminopyridin (DAP) groups were immobilized onto thymine-modified nanofibers through thymine-DAP base pairing in CCl_4. This research introduces new insights for developing biological/inorganic hybrid nanomaterials.

Peptide nanofibers were used as nano-templates not only for inorganic materials, but also for 1D-protein array. The biotinylated peptides were incorporated into the nanofibers co-assembled with non-biotynylated peptides, allowing regular immobilization of streptavidin (SAv) onto the nanofibers [76, 77]. Since the SAv has four biotin-binding sites, the additional and versatile functionalization of SAv-immobilized nanofiber would be possible by using various biotin derivatives. Our group successfully accomplished the introduction of chemically synthesized biotin-terminated PNIPAM onto the biotin-functionalized nanofiber through biotin-avidin complexation at pH 9 (Fig. 23). The resultant PNIPAM/SAv-immolized nanofibers showed thermo-response based on the character of surface-bound PNIPAM segment. Figure 23C shows the TEM images of PNIPAM/SAv-immolized nanofibers.

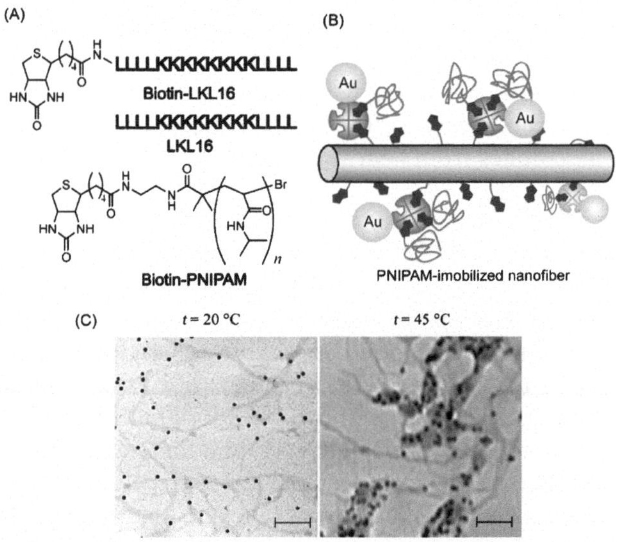

Fig. 23 **A** Chemical structures of biotin-terminated LKL16 and PNIPAM used as building units for the thermo-responsive nanofiber architecture. **B** Schematic illustration of the nano-organization of PNIPAM onto biotin-LKL16/LKL16-mixed nanofiber template through biotin-SAv complexation. **C** TEM images of PNIPAM/SAv-immolized nanofibers prepared onto carbon-coated copper grids at 20 °C (below LCST of PNIPAM) or 45 °C (above LCST of PNIPAM)

TEM samples were prepared onto carbon-coated copper grids at 20 °C (below LCST of PNIPAM) or 45 °C (above LCST of PNIPAM), and Au nanoparticle-modified SAv was used to make SAv visible under TEM observation. At below LCST of PNIPAM segments (20 °C), the SAv located on nanofiber discretely. On the other hand, at above LCST of PNIPAM segments, the nanofibers aggregated with each other due to dehaydrated PNIPAM–PNIPAM interaction.

Thus, the engineered peptide nanofibers can be applied to develop nano-arrays of various chemical and biological molecules.

3 Helix-Based Self-Assembled Nanofibers

Peptide self-assembly as a tool for nanofabrication has been given a lot of attention in recent years. As we described above, such assemblies have potential as novel nanobiomaterials. Considerable work has been done to fabricate fibrous peptide assemblies by using β-sheet-forming peptides (see Sect. 2). Relatively less has been done with α-helix-based nanofiber architectures. However, the potential of using helical peptides for the construction of well-defined peptide nanofibers remains great,considering the fact that fibrils derived from α-helical coiled-coil motifs exist widely in native biological systems, such as in an extracellular matrix. Collagen triple helix structures will also be useful as structural motif for nano-organization. Thus, molecularly engineered helical peptides are also an attractive target as self-assembling building units for structurally-regulated nanofiber formation. In this section, we focus on the recent works for the design concepts and characterization of fibrous nano-assemblies from helical peptides.

3.1 Coiled-Coil Motif

Coiled-coil helix is one of the most common folding motifs in proteins. This conformation has been found in a wide variety of proteins including structural proteins and leucine zipper proteins of transcription factors. The former, for example, α-keratin is known to adopt coiled-coil helices, which pack parallel orientations with straight axes, and the axes of individual helices pursuing a helical course to form multi-stranded fibers [78, 79]. In the later case, GCN4, a special type of coiled-coil protein that belongs to a class of transcription factors, regulates the expression of many different genes in both simple and complex organisms [80]. One structural feature of coiled-coil is a bundle of two- or more right-handed amphiphilic α-helices wrapping around each other into a slightly left-handed superhelix. The typical amino acid sequence of a coiled-coil is a seven amino acid repetition, $(a, b, c, d, e, f, g)_n$ where a and d amino acids are hydrophobic residues, and other residues are usually polar ones. The hydrophobic interaction between residues *a* and *d* at the hydrophobic core, which is formed by the side-by-side packing of two α-helices, constitutes the main force for stabilizing the super coiled helices (e.g., Fig. 24C). Ionic residues are often used at positions *e* and *g* in the designed coiled-coil peptides. The stabilization of the supercoiled helix is enhanced or diminished by positioning attractive or repulsive charged groups at *e* and *g* positions. It is also possible to design polypeptides that have the propensity to form parallel and anti-parallel orientation based on charge-charge interactions at the *e* and *g* positions on adjacent helices. A number of studies have been performed on the use of the coiled-coil motif as a functional peptide-based material [81–83]. In the course of such research, some of fabrications of designed α-helical coiled-coil-based nanofibers have been accomplished.

Woolfson et al. have produced nanofibers with a design based on a dimeric coiled-coil structure [84–89]. They successfully fabricated self-associating coiled-coil nanofibers by using a "sticky-end" design, similar to that found in DNA assembly. Figure 24 shows their design concept and sequences of the

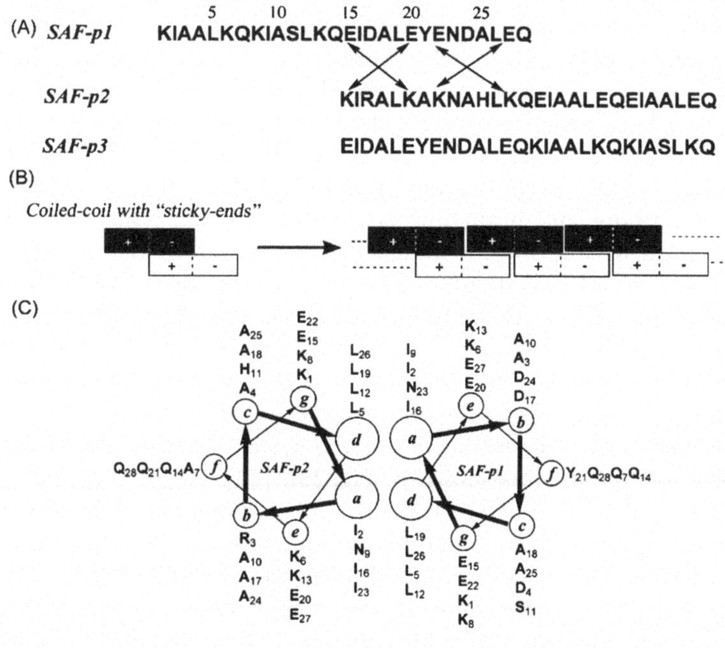

Fig. 24 Coiled-coil-based self-assembling peptide system designed by Woolfson's group. **A** Amino acid sequences of self-assembing fiber (SAF) peptides (SAF-p1, -p2 and -p3). **B** Concept for a sticky-end assembly process. Complementary charges in companion peptides direct the formation of staggered, parallel heterodimers; the resultant sticky-end are also complementary and promote longitudinal association into extranded nanofibers. **C** Helical wheel representation of coiled-coil conformation. (Adapted from [85])

first generation self-assembling fiber (SAF) peptides [85]. They first designed two 28-residue SAF peptides (SAF-p1 and SAF-p2) to form extended fibers when mixed by folding them into a staggered heterodimer with overhanging ends (Fig. 24A). Each peptide comprised of a heptad repeat (*a*, *b*, *c*, *d*, *e*, *f*, *g*) with Ile at *a* and Leu at *d* to form coiled-coil dimers. In order to stabilize the staggered dimmer with sticky-ends, oppositely charged Glu and Lys residues were incorporated at positions *e* and *g*. Interestingly, Asn residues, which preferentially form a hydrogen bonding pair with each other in coiled-coil dimer, were introduced to cement the above conformation and to favor parallel structures. For comparison, SAF-p3 was also designed as a circular permutation of SAF-p1 with residues 15–18 swapped for residues 1–14. CD spectra clearly showed a coiled-coil helix formation with negative bands at 222 nm and 208 nm when the SAF-p1 and SAF-p2 were mixed at 100 μM, pH 7 and 5 °C. Prolonged incubation for 1 h caused the spectral change: the signal at 222 nm doubled in intensity while the signal at 208 nm remained almost unchanged. In parallel with such structural change, TEM images demonstrated that the matured nanofiber possessed a width of ca. 40–50 nm and a length > 10 μm. In contrast, such nanofiber formation was not observed in the mixed systems of SAF-p2 and SAF-p3, which formed heterodimers with "blunt-ends". Interestingly, the observed nanofibers were thicker than the average diameter of two-stranded coiled-coil (ca. 2 nm). Therefore, the author concluded that the designed two-stranded coiled-coil nanofibers associated laterally into high-order assemblies. In fact, this lateral assembly was also confirmed by using fluorescently labeled SAF-p1 and SAF-p2 [86]. Woolfson's group also developed second-generation SAF peptides that enable the introduction of kinking and branching structures into the nanofibers [87–89]. They first tried to add non-linear specific peptides, SAF-C2^N and SAF-D2^C, into the above mentioned standard SAF peptides mixture [88, 89]. Figure 25A shows the primary structures of newly designed specific SAF building units. The amount of kinking and branching was changed by the type and concentration of the specific peptides, but this SAF system was more susceptible to kinking [87]. In order to introduce the branching structure into coiled-coil nanofiber, they next designed a mixed system of standard SAF peptide and "T-shaped" peptide as shown in Fig. 25B [89]. For SAF-p1 and SAF-p2 alone, TEM and SEM analyses revealed linear nanofibers (ca. 50 nm thick and a length > μm) without junctions or branches. On the other hand, when T-shaped SAF was co-assembled with SAF-p1 and SAF-p2 (100 μM each peptides, pH 7.0, 20 °C), branching points were produced in nanofibers. The authors claimed that one problem with this system was that the branching structure appeared not to be completely tolerated by the nanofibers; in which ∼ 30% of the nanofibers had branches.

Some other excellent works concerning coiled-coil helix-based self-assembling nanofibers have been accomplished in last 5–10 years [90–97]. Ogihara et al. reported another strategy to create one-dimensional organi-

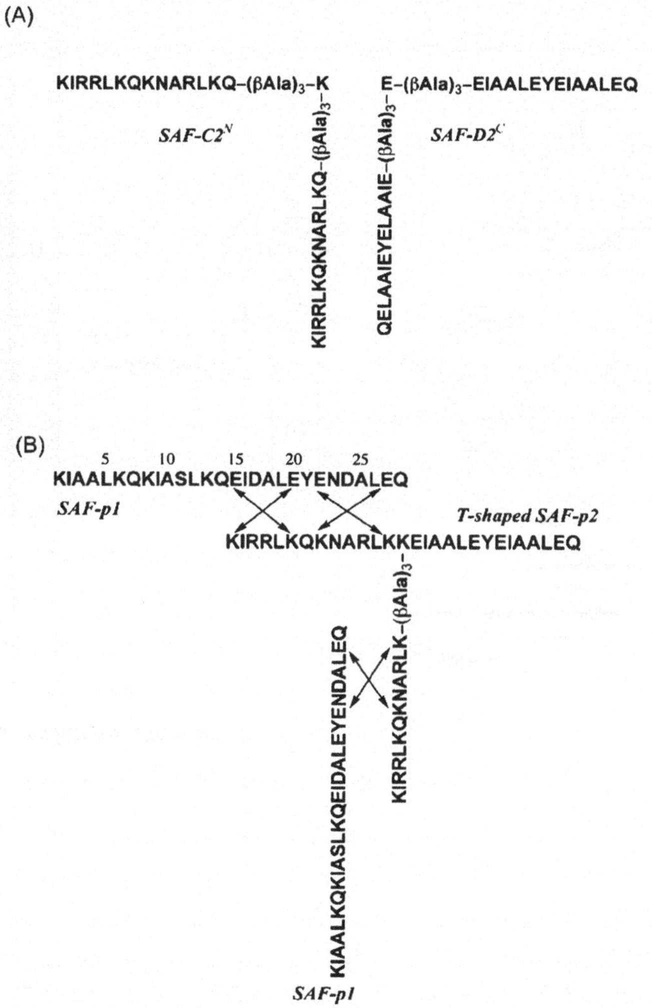

Fig. 25 Molecular structures of the 2nd-generation of SAF and T-shaped SAF building blocks developed by Woolfson's group

zation of coiled-coil helix by controlling the topology [91]. They employed three-dimensional domain-swapped protein, consisting of a three coiled-coil helix bundle with an up-down-down topology (Fig. 26). TEM, CD, and FTIR clearly showed the long fibrous aggregates with a width of 40–70 nm, and coiled-coil helical conformation at both acidic and neutral pHs. The fibrils were considered to be composed of several protofibrils, which corresponded to the thickness of one to three triple-stranded α-helical bundles. Potekhin et al. proposed a structural model for the rational design of α-helical nanofibers with five-stranded coiled-coil through self-assembly of 34-mer oligopeptide

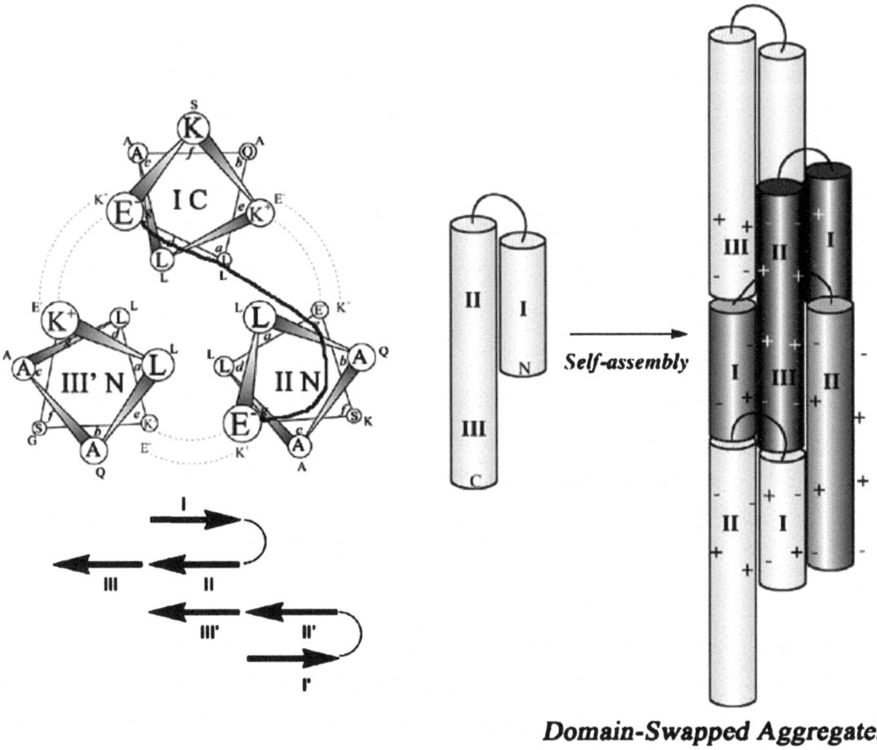

Fig. 26 Design of a domain-swapped open aggregate (DSAg) of Mon 2 with up-down-down topology. The designed *e*–*g* salt bridges stabilizing the conformation for DSAg (*right figure*). Helical wheel diagram and amino acid sequence of DSAg (*left figure*). (Adapted from [91])

(QLAREL(QQLAREL)$_4$) [92]. On the basis of their model, an oligopeptide with a heptad repeat sequence can self-assemble into an n-stranded rope (n: degree of oligomerization of the helix bundle) and the shifts Δl of the adjacent α-helices should be equal to a value divisible by one heptad (Fig. 27). On the other hand, the staggered helices of the n-stranded rope complete one turn without overlapping, if the Δl shift is more than $(L + \delta l)/n$, where L = amino acid residues, δl = residue equivalents (usually 1 or 0, for single space for head-to-tail packing of α-helices). The shift of more than $(L + \delta l)/n$, may be unfavorable because it leads to the head-to-tail gap and exposure of some hydrophobic amino acids at a and d positions to the solvent. Therefore, they concluded that the most favorable minimal length of the α-helix for an n-stranded rope was $(n \times 7 - 1)$ residues, i.e., 20, 27, and 34 residues for three-, four-, and five-stranded fibrils, respectively, as shown in Fig. 27. Zimenkov et al. constituted an experimental verification of the Potekhin model. They observed the formation of long aspect-ratio he-

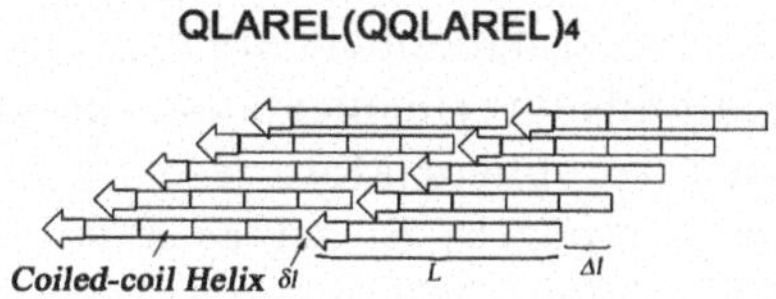

Fig. 27 Schematic structural model of the self-assembly of a 34-mer α-helical peptide (QLAREL(QQLAREL)$_4$) into a five-stranded coiled-coil fiber reported by Potekihin et al. (Adapted from [92])

lical nanofiber with uniform diameter ranged from ca. 20–50 nm, in which the fundamental structural subunit corresponded to a dimeric coiled-coil associated laterally by thermal annealing of 42-mers peptide (YZ1; Ac-EIAQLEKEIQALEKENAQLEKKIQALRYKIAQLREKNQALRE-CONH$_2$) under controlled conditions [93]. Reversible assembly of helical filaments was also reported by Frost et al. from de novo designed 18-mers short peptide (KIA13: AKAIAAIAKAIAAIAKAG) in buffer at pH 7 [94, 95]. From the CD spectroscopy, turbidity and in-situ TM-AFM measurements, KIA13 peptide was found to form insoluble nanofibers with coiled-coil association at high NaCl concentrations ($>$ 850 mM), soluble fibers at intermediate NaCl concentrations (150-850 mM), and exist in an unfolded state at low NaCl concentrations. Reducing the NaCl concentration completely reversed the above process. In this molecular system, hydrophobic interactions were used to drive assembly into nanofibers. Similarly, Fairman et al. also pointed out that using the hydrophobic effect to drive self-assembly circumvents problems of uncontrolled self-assembly using electrostatics as a mode of self-assembly [96]. They designed a peptide, CpA, consisting of two GCN4-like sequences separated by two Ala-residues. The Ala-Ala insertion introduced a phase shift in the heptad repeat, generating two hydrophobic faces oriented at 200° with respect to each other on the α-helix, and, as a result, a CpA dimer with hydrophobic "sticky ends" was produced to initiate a well-regulated polymerization into nanofilaments and nanoropes (Fig. 28).

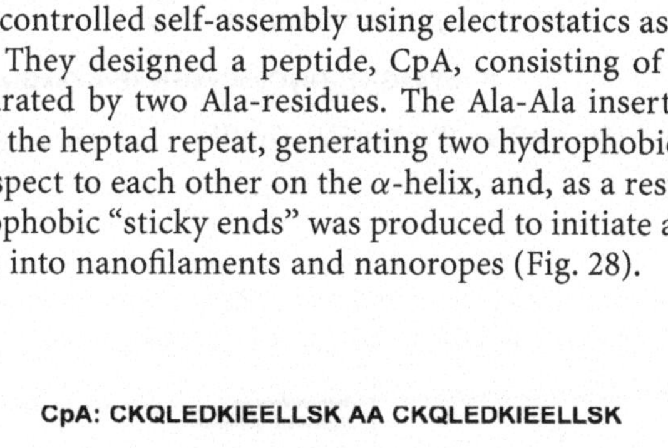

Fig. 28 Molecular model of a self-asembled polymer from the CpA. The two-alanine insertion results in a phase-shift of ca. 200° of the C-terminus hydrophobic surface, relative to the N-terminus hydrophobic surface, leading to a CpA dimer with hydrophobic sticky-end. (Adapted from [96])

These coiled-coil helix-based nanofibers are expected to be used as nano-templates and nano-scaffolds in chemical and biological fields, as also expected in β-sheet-based nanofibers. More recently, in fact, coiled-coil self-assembling nanofiber was reported to act as a template for polypeptide synthesis [97].

3.2
Another Class of Helix Nanofibers

Another type of helix-based nanofiber using the self-assembly "bottom-up" approach was also fabricated using collagen-like triple helix foldamer as a building unit. Collagen is the major protein constituent of diverse tissues such as bone, skin, and cornea [98]. Its high natural abundance and intrinsic plasticity have accelerated the development of collagen as a biometerial [99, 100]. Collagen is a rod-like macromolecule with ca. 280 nm length, a diameter of 1.2 nm, and consists of three strands that wind into a coaxial helix (triple helix) [101–103]. Each strand has the repeating sequence X_{aa}-Y_{aa}-Gly with the most typical triplet being Gly-Pro-Hyp. Various chemically synthesized collagen-mimetics have been reported, but such triple helices are limited to the length (< 10 nm) in many cases. To overcome such a problem, Raines et al. applied the "sticky-end" strategy to a collagen self-assembling molecular system [104]. Figure 29 shows the primary structures

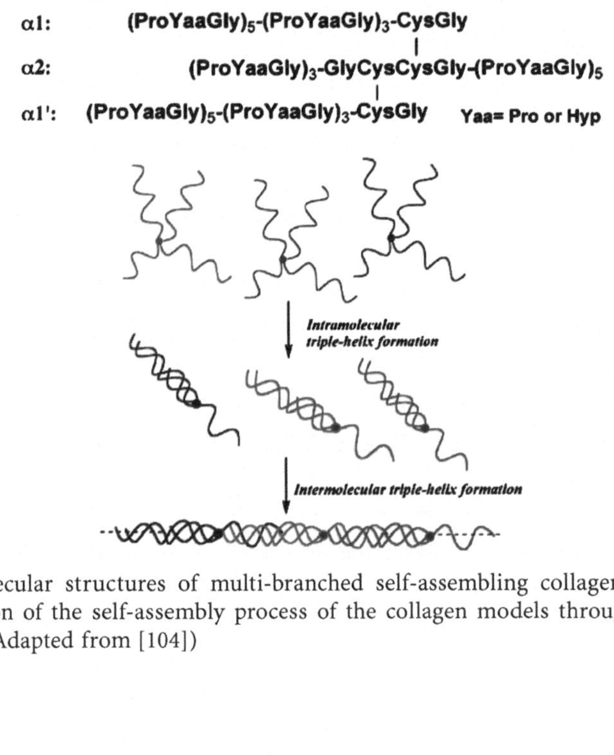

Fig. 29 Molecular structures of multi-branched self-assembling collagen-models, and representation of the self-assembly process of the collagen models through triple-helix formation. (Adapted from [104])

of their collagen-mimetics with sticky-end. The $(ProHypGly)_3$ segments in the two identical strands ($\alpha1$ and $\alpha1'$) and third strand ($\alpha2$) form an intramolecular triple helix. The data of CD, DLS, ultracentrifugation, AFM, and TEM measurements indicated the formation of natural collagen-like long nanofibers (> 400 nm) through the intermolecular assembly of the overhanging $(ProHypGly)_5$ segments (Fig. 29).

Our research group also fabricated a peculiar α-helix-based nanofiber by the precise design of helix–helix interaction. Figure 30 shows the amino acid sequence and structural model of the amphiphilic α-helical 18-mers peptide used as a building unit for self-assembly. The designed peptide contains the following sequence: AEAQAQAEAARAQAQARA. Due to the high Ala content, the peptide is helical, and provides hydrophobic and hydrophilic

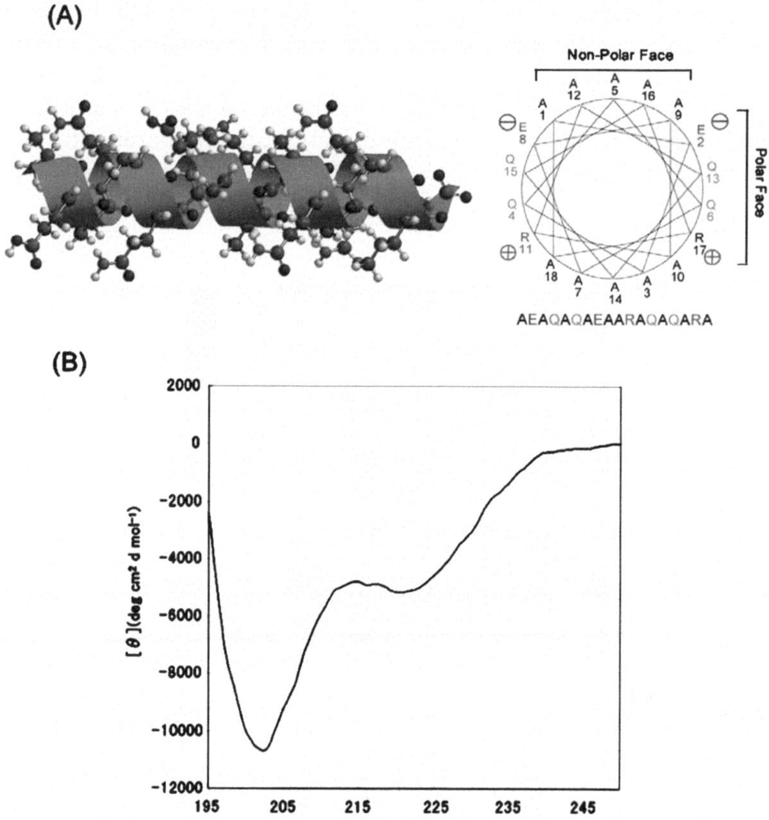

Fig. 30 **A** Structural model of de novo designed α-helical 18-mers oligopeptide with distinct polar and non-polar faces, AEAQAQAEAARAQAQARA. **B** CD spectrum of the peptide in Tris/HCl buffer at pH 7.0. The CD spectrum was time-independent within the period of 24 h. [peptide] = 40 μM

surfaces. Oppositely charged Arg and Glu residues were incorporated to cement the helix–helix interaction with each other. The Gln residues, as well as the Arg and Glu residues, promoted solubility in water (Fig. 30A,B). CD spectra demonstrated that the peptide formed α-helix structure in a buffer in both cases at pH 3.1 and pH 7.0 (Fig. 30C), although the helix content at pH 7.0 was slightly higher than that at pH 3.1. It was found from an AFM study that the α-helical peptide self-assembled into tape-like thin and straight nanofibers with a height of 0.8–1.0 nm at pH 7.0 (Fig. 31A). The observed height corresponds well to the diameter of the α-helix. Therefore, these α-helix nanofibers probably consist of a helix monolayer, in which the peptide helices interact with each other in two-dimensions, through the hydrophobic interaction between Ala residues and electrostatic interaction between Glu and Arg residues. In fact, such nanofiber formation was not observed at the condition of pH 3.1, at which the negative charges present on the Glu residues were shielded or had disappeared. Moreover, the helix-macrodipole interaction (namely, the head-to-tail antiparallel orientation of the helices, which is energetically more favorable) may also contribute to the stability of the tape-shaped, high-order structure. The dipole moments of amino acid in an α-helical peptide align in the same direction, nearly parallel to the helix axis, then the resulting macroscopic dipole generates an electrostatic poten-

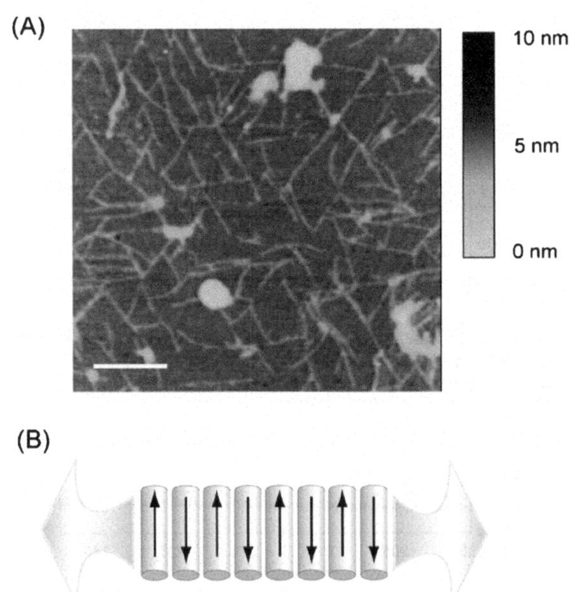

Fig. 31 **A** Tapping-mode AFM image of AEAQAQAEAARAQAQARA obtained at pH 7.0. [peptide] = 40 μM. *z*-Scale: 10 nm. **B** Structural model of the observed α-helical nanofibers

tial, directed from the N-terminus to the C-terminus [21]. We reported direct evidence for helix-helix macrodipole interaction by exploring an attractive interaction between the disulfide-modified poly(L-glutamic acid)(PLGA) self-assembled monolayer on gold and redox active PLGA derivatives as guest helices [105–107]. Thus, this macrodipole interaction, as well as the hydrophobic interaction and ionic pairing, should be one of the useful mediators used to construct the high-order structure of α-helix-based nanofiber (Fig. 31B).

Thus, these works offer intriguing possibilities of precisely designed synthetic helical peptides for the construction of functional nano-scaled materials, including new types of nanofibers.

4
Summary

Molecular self-assembly is a good candidate for a breakthrough technology that enables the fabrication of complicated three-dimensional objects from the bottom-up in the region scaled between nanometers and submicrons, which is difficult to accomplish by the conventional top-down approaches. In this paper, we reviewed some recent powerful strategies for fabricating various supramolecular nanoarchitectures, focusing especially on nanofiber architectures, through controlled self-assembly of artificial peptide building blocks. Interest in polypeptides as building blocks is driven by their ability to form well-regulated high-order structures such as α-helix and β-sheet, and to hierarchically self-organize into 3D-nanostructures. Moreover, such conformations of artificial peptides can fundamentally be designed by selecting the amino acid sequence and environmental conditions, and as a result, various types of peptide nanofibers have been constructed. The possible applications for such peptide-based nanofibers as a new class of soft-materials are wide, from the field of biomedical engineering, such as tissue engineering and biomineralization, to the nanotechnology field for engineering devices, sensors, and electronic nanocomponents. Investigating the self-assembly of synthetic peptides is also likely to make an important contribution to our understanding of the mechanism of protein folding and aggregation.

Acknowledgements The authors are grateful for the contributions of those people who are cited in the references of our research and to the members of our laboratory. The financial support from the grant-in-aid for scientific research (No. 17710104, 19710099) from the Ministry of Education, Culture, Sports, Science, and Technology, the Academic Frontier Research Project on "New Frontier of Biomedical Engineering Research" of Doshisha University, and the Ministry of Education, Culture, Sports, Science and Technology are also gratefully acknowledged.

References

1. Sipe JD (1994) Crit Rev Clin Lab Sci 31:325
2. Lansbury PT (1999) Proc Natl Acad Sci USA 96:3342
3. Dobson CM (1999) Trends Biochem Sci 24:329
4. Serpell LC (2000) Biochim Biophys Acta 1502:16
5. Selkoe DJ (1996) J Biol Chem 271:18295
6. Prusiner SB (1991) Science 252:1515
7. Kelly JW (1998) Curr Opin Struc Biol 8:101
8. Rochet JC, Lansbury PT (2000) Curr Opin Struc Biol 10:60
9. Sunde M, Blake CCF (1997) Adv Protein Chem 50:123
10. Chiti F, Webster P, Taddei N, Clark A, Stefani M, Ramponi G, Dobson CM (1999) Proc Natl Acad Sci USA 96:3590
11. Fandrich M, Fletcher MA, Dobson CM (2001) Nature 410:165
12. Fezoui Y, Hartley DM, Walsh DM, Selkoe DJ, Osterhout JJ, Teplow DB (2000) Nat Struct Biol 7:1095
13. Koga T, Higuchi M, Kinoshita T, Higashi N (2006) Chem Eur J 12:1360
14. Shamovsky JL, Ross GM, Riopelle RJ (2000) J Phys Chem B 104:11296
15. Aggeli A, Nykova IA, Bell M, Harding R, Carrick L, McLeish TCB, Semenov AN, Boden N (2001) Proc Natl Acad Sci USA 98:11857
16. Fishwick CWG, Beevers AJ, Carrick L, Whitehouse CD, Aggeli A, Boden N (2003) Nano Lett 3:1475
17. Aggeli A, Bell M, Carrick L, Fishwick CWG, Harding R, Mawer PJ, Radford SE, Strong AE, Boden N (2003) J Am Chem Soc 125:9619
18. Jemenez JL, Nettleton EJ, Bouchard M, Robinson CV, Dobson CM (2002) Proc Natl Acad Sci USA 99:9196
19. Koga T, Matsuoka M, Higashi N (2005) J Am Chem Soc 127:17596
20. Yokoi H, Kinoshita T, Zhang S (2005) Proc Natl Acad Sci USA 102:8414
21. Wada A (1976) Adv Biophys 9:1–63
22. Zhang S, Holmes T, Lockshin C, Rich A (1993) Proc Natl Acad Sci USA 90:3334
23. Fung SY, Keyes C, Duhamel J, Chen P (2003) Biophys J 85:537
24. Hong Y, Legge RL, Zhang S, Chen P (2003) Biomacromolecules 4:1433
25. Jun S, Hong Y, Imamura H, Ha B-Y, Bechhoefer J, Chen P (2004) Biophys J 87:1249
26. Marini DM, Hwang W, Lauffenburger DA, Zhang S, Kamm RD (2002) Nano Lett 2:295
27. Hwang W, Marini DM, Kamm RD, Zhang S (2003) J Chem Phys 118:389
28. Caplan MR, Schwartzfarb EM, Zhang S, Kamm RD, Lauffenburger DA (2002) Biomaterials 23:219
29. Goeden-Wood NL, Keasling JD, Muller SJ (2003) Macromolecules 36:2932
30. Wang K, Keasling JD, Muller SJ (2005) Int J Biol Macromol 36:232
31. Lamm MS, Rajagopal K, Schneider JP, Pochan DJ (2005) J Am Chem Soc 127:16692
32. Matsumura S, Uemura S, Mihara H (2004) Chem Eur J 10:2789
33. Hartgerink JD, Beniash E, Stupp SI (2001) Science 294:1684
34. Hartgerink JD, Beniash E, Stupp SI (2002) Proc Natl Acad Sci USA 99:5133
35. Silva GA, Czeisler C, Niece KL, Beniash E, Harrington DA, Kessler JA, Stupp SI (2004) Science 303:1352
36. Paramonov SE, Jun H-W, Hartgerink JD (2006) J Am Chem Soc 128:7291
37. Niece KL, Hartgerink JD, Donner JJJM, Stupp SI (2003) J Am Chem Soc 125:7146
38. Behanna HA, Donners JJJM, Gordon AC, Stupp SI (2005) J Am Chem Soc 127:1193

39. Lashuel HA, LaBrenz SR, Woo L, Serpell LC, Kelly JW (2000) J Am Chem Soc 122:5262
40. Takahashi Y, Ueno A, Mihara H (1998) Chem Eur J 4:2475
41. Takahashi Y, Ueno A, Mihara H (2002) ChemBioChem 3:637
42. Yang Z, Liang G, Ma M, Gao Y, Xu B (2007) J Mater Chem 17:850
43. Collier JH, Hu B-H, Ruberti JW, Zhang J, Shum P, Thompson DH, Messeersmith PB (2001) J Am Chem Soc 123:9463
44. Bosques CJ, Imperialli B (2003) J Am Chem Soc 125:7530
45. Koga T, Kitamura K, Higashi N (2006) Chem Commun 4897
46. Koga T, Kitamura K, Higashi N (2007) Proc Am Pep Soc, in press
47. Mutter M, Chandravarkar A, Boyat C, Lopez J, Santos SD, Mandal B, Mimna R, Murat K, Patiny L, Saucede L, Tuchschere G (2004) Angew Chem Int Ed 43:4172
48. Matsumura S, Uemura S, Mihara H (2006) Supramolec Chem 18:397
49. Koga T, Ushirogochi M, Higashi N (2007) Polym J 39:16
50. Koga T, Watanabe T, Higashi N (2008) J Nanosci Nanotech, in press
51. Burkoth TS, Benzinger TLS, Urban V, Morgan DM, Gregory DM, Thiyagarajan P, Botto RE, Meredith SC, Lynn DG (2000) J Am Chem Soc 122:7883
52. Collier J, Messersmith PB (2004) Adv Mater 16:907
53. Eckhardt D, Groenewolt M, Krause E, Börner HG (2005) Chem Commun 2814
54. ten Cate MGJ, Severin N, Börner HG (2006) Macromolecules 39:7831
55. Koga T, Taguchi K, Kinoshita T, Higuchi M (2002) Chem Commun 1126
56. Koga T, Taguchi K, Kobuke Y, Kinoshita T, Higuchi M (2003) Chem Eur J 9:1146
57. Pauling L, Corey RB (1951) Proc Natl Acad Sci USA 37:729
58. Mori T, Yasutake S, Inoue H, Minagawa K, Tanaka M, Niidome T, Katayama Y (2007) Biomacromolecules 8:318
59. Matsumoto A, Sada K, Tashiro K, Miyake M, Tsubouchi T, Tanaka T, Odani T, Nagahama S, Tanaka T, Inoue K, Saragai S, Nakamoto S (2002) Angew Chem Int Ed 41:2502
60. Jahnke E, Lieberwirth I, Severin N, Rabe JP, Frauenrath H (2006) Angew Chem Int Ed 45:5383
61. Ciuet J, Samuel JDJS, Kopyshev A, Santer S, Biesalski M (2005) Angew Chem Int Ed 44:3297
62. Zhang S (2003) Nat Biotech 21:1171
63. Zhang S, Zhao X (2004) J Mater Chem 14:2082
64. Weiner S, Traub W (1986) FEBS Lett 16:224
65. Weiner S, Addadi L (1997) J Mater Chem 7:689
66. Pierschbacher MD, Ruoslahti KP (1987) J Biol Chem 262:17294
67. Hosseinkhani H, Hosseinkhani M, Tian F, Kobayashi H, Tabata Y (2006) Biomaterials 27:4079
68. Harrington DA, Cheng EY, Guler MO, Lee LK, Donovan JL, Claussen RC, Stupp SI (2006) J Biomed Mater A 78:157
69. Kisiday J, Jin M, Kurz B, Hung H, Semino C, Zhang S, Grodzinsky AJ (2002) Proc Natl Acad Sci USA 99:9996
70. Semino CE, Merok JR, Crane GG, Panagiotakos G, Zhang S (2003) Differentiation 71:262
71. Kretsinger JK, Haines LA, Ozbas B, Pochan DJ, Schneider JP (2005) Biomaterials 26:5177
72. Reches M, Gazit E (2003) Science 300:625
73. Patolsky F, Weizmann Y, Willner I (2004) Nat Mater 3:692
74. Fu X, Wang Y, Huang L, Sha Y, Gui L, Lai L, Tang Y (2003) Adv Mater 15:902
75. Li L-S, Stupp SI (2005) Angew Chem Int Ed 44:1833

76. Kodama H, Matsumura S, Yamashita T, Mihara H (2004) Chem Commun 2876
77. Koga T, Murakami R, Higashi N (2007) Trans Mater Res Soc Jpn 32:371
78. Astbury WT, Haggith JW (1953) Biochim Biophys Acta 10:483
79. Pauling L, Corey RB (1953) Nature 171:59
80. O'Shea EK, Klemm JD, Kim PS, Alber T (1991) Science 254:539
81. Petka WA, Harden JL, McGrath KP, Wirtz D, Tirrell DA (1998) Science 281:389
82. Tang A, Wang C, Stewart RJ, Kopecek J (2001) J Controlled Release 72:57
83. Vandermeulen GWM, Tziatzios C, Klok H-A (2003) Macromolecules 36:4107
84. MacPhee CE, Woolsfon DN (2004) Curr Opin Solid State Mater Sci 8:141
85. Pandya MJ, Spooner GM, Sunde M, Thorpe JR, Rodger A, Woolfson DN (2000) Biochemistry 39:8728
86. Smith AM, Acquah SFA, Bone N, Kroto HW, Ryadov MG, Stevens MSP, Walton DRM, Woolfson DN (2005) Angew Chem Int Ed 44:325
87. Ryadnov MG, Woolfson DN (2003) Nat Mater 2:329
88. Ryadnov MG, Woolfson DN (2005) J Am Chem Soc 127:12407
89. Ryadnov MG, Woolfson DN (2003) Angew Chem Int Ed 42:3021
90. Padilla JE, Colovos C, Yeates TO (2001) Proc Natl Acad Sci USA 98:2217
91. Ogihara NL, Ghirlanda G, Bryson JW, Gingery M, DeGrado WF (2001) Proc Natl Acad Sci USA 98:1404
92. Potekhin SA, Melnik TN, Popov V, Lanina NF, Vazina AA, Rigler P, Verdini AS, Corradin G, Kajava AV (2001) Chem Biol 8:1025
93. Zimenkov Y, Conticello VP, Guo L, Thiyagarajan P (2004) Tetrahedron 60:7237
94. Boon CL, Frost D, Chakrabartty A (2004) Biopolymers 76:244
95. Frost DW, Yip CM, Chakrabartty A (2005) Biopolymers 80:26
96. Wagner DE, Phillips CL, Ali WM, Nybakken GE, Grawford ED, Schwab AD, Smith WF, Fairman R (2005) Proc Natl Acad Sci USA 102:12656
97. Ryadnov MG, Woolfson DN (2007) J Am Chem Soc 129:14074
98. Hulmes DJS, Miller A (1979) Nature 282:878
99. Ramshaw JAM, Werkmeister JA, Glattauer V (1996) Biotechnol Genet Eng Rev 13:335
100. Lee C, Singla A, Lee Y (2001) Int J Pharm 221:1
101. Hall CE, Doty PJ (1958) J Am Chem Soc 80:1269
102. Ramachandran GN, Kartha G (1954) Nature 174:269
103. Rich A, Crik FH (1961) J Mol Biol 3:483
104. Kotch FW, Raines R (2006) Proc Natl Acad Sci USA 103:3028
105. Niwa M, Morikawa M, Higashi N (1999) Langmuir 15:5088
106. Niwa M, Morikawa M, Higashi N (2000) Angew Chem Int Ed 39:960
107. Niwa M, Kuwagaki Y, Yamaguchi S, Higashi N (2003) Angew Chem Int Ed 42:1839

Adv Polym Sci (2008) 219: 69–106
DOI 10.1007/12_2008_138

Published online: 5 April 2008

Self-Assembled Nanofibers and Related Nanostructures from Molecular Rods

Byoung-Ki Cho[1] (✉) · Ho-Joong Kim[2] · Yeon-Wook Chung[1] ·
Byung-Ill Lee[1] · Myongsoo Lee[2] (✉)

[1]Department of Chemistry and Institute of Nanosensor and Biotechnology,
Dankook University, 448-701 Gyeonggi, Korea
chobk@dankook.ac.kr

[2]Center for Supramolecular Nano-Assembly and Department of Chemistry,
Yonsei University, 120-749 Seoul, Korea
mslee@yonsei.ac.kr

1 **Introduction** 70

2 **One-Dimensional Assemblies of Molecular Rods in Solutions** 72
2.1 Rod–Coil Molecules 72
2.2 Macrocyclic Molecules 77
2.3 Dendron–Rod–Coil Molecules 80
2.4 Dumbbell-Shaped Molecules 82
2.5 Wedge-Type Molecules 87
2.6 Conjugate Rods with Lateral Chains 89
2.7 Biomimetic or Bioconjugate Amphiphilic Rods 96

3 **Conclusions and Outlook** 102

References 104

Abstract This review introduces recent research in the area of self-assembled nanostructures from rod molecules in solutions; in particular one-dimensional nano-aggregates such as fibers, ribbons, tubules are highlighted. Self-assembled nanostructures are well-known to be strongly dependent upon the architecture of molecular building blocks; thus, in this review we discuss assembling behavior and the related functions associated with molecular shapes such as simple rod–coils, macrocycles, dendron–rod–coils, dumbbells, wedges and conjugated rods with lateral chains. In addition, biomimetic or bioconjugate amphiphilic rod systems are described.

Keywords Cylindrical micelles · One-dimensional nanostructures · Molecular rod · Nanofibers · Self-assembly · Supramolecular objects

Abbreviations

AFM Atomic force microscopy
CD Circular dichroism
DLS Dynamic light scattering
DRC Dendron–rod–coil
DSC Differential scanning calorimetry

ED	Electron diffraction
EHMA	2-Ethylhexyl methacrylate
EMA	Ethyl methacrylate
FESEM	Field emission scanning electron microscopy
FRET	Fluorescence resonance energy transfer
HOMO	Highest occupied molecular orbital
HP	Hexa-*para*-phenylene
LUMO	Lowest unoccupied molecular orbital
OP	Oligo(*para*-phenylene)
OPV	Oligo(*para*-phenylene vinylene)
OPE	Oligo(*para*-phenylene ethynylene)
PEO	Poly(ethylene oxide)
PPO	Poly(propylene oxide)
SANS	Small-angle neutron scattering
SAXS	Small-angle X-ray scattering
SEM	Scanning electron microscopy
STM	Scanning tunneling microscopy
TEM	Transmission electron microscopy
WAXS	Wide-angle X-ray scattering
XRD	X-ray diffraction

1 Introduction

Creation of supramolecular architectures that have a well-defined shape and size has been one of the major research areas of materials science, nanochemistry, and biomimetic or bioinspired chemistry. Fine tuning of supramolecules is particularly important due to the operative functions. Self-assembly of molecules via secondary forces including hydrogen bonding, electrostatic, van der Waals interactions, and the hydrophobic effect is a powerful tool to develop desired supramolecular functional materials and artificial biomaterials [1–5]. Among a variety of self-assembling soft materials, e.g. coil–coil block copolymers or small molecular surfactants, in recent years, molecular rods containing flexible coils have received plenty of attention as self-assemblers [6]. The peculiar self-assembling power of the molecular rod system which stimulates researchers' interest is the distinct conformational and shape contrasts between rigid rod and flexible coil blocks, as well as amphiphilicity from designed block compositions. Via these driving forces, rod molecules with coil segments can show microphase-separation into ordered structures. In addition, the enhanced segregation leads to assembled structure with relatively small dimensions (a few nanometers), compared to conventional coil–coil block copolymers. Many researchers have studied supramolecular structures using diverse molecular rod blocks such as helical peptides [7, 8], isocynates [9], mesogenic phenyl units [10] and rigid conjugate aromatic units [11, 12]. Researchers have found that observed morphologies are determined by the cooperative effects of several molecular parameters such as volume fraction, rod anisotropy, molecular structural topology and coil-cross sectional area [13–15].

Meanwhile, block molecules may associate to form aggregates when they are exposed to a selective solvent that is a good solvent for one block and a poor solvent for the other block. The solvent-assisted supramolecular structures are strongly dependent upon solvent polarity, block composition/architecture, temperature and molecular weight, which substantially determine solvent–block interactions [16–19]. Several self-assembling molecular systems such as glycolipids [20], lipid–peptides [21, 22], polymer–metal molecular hybrids [23] and coil–coil block copolymers [24] have been studied extensively and show a wide variety of aggregate structures including spheres, cylinders, vesicles and toroids. Several review articles have discussed solution behavior of the above molecular building blocks [25, 26]. A newly emerging molecular candidate, i.e. rod building blocks bearing flexible coils, may self-assemble into ordered nanostructures as long as the molecular structures are rationally designed to be endowed with self-assembling features. Although diverse morphologies are expected, one-dimensional nanostructures like fibers, tubules and ribbons with well-defined shape and size are one of the most important subjects in material or bio-related chemistry. For example, a 1-D shape with micro-length scales can be used for electro-active supramolecular wires for the development of nanodevices, scaffolding units of stable gel materials, structure-directing agents for shape-specific mineralization etc. [27–31]. From the beginning of the rod system, rod blocks have employed non-mesogenic units such as polyferrocenes [32] polypeptides [33] and polyisocyanides [34, 35] where 1-D hollow nanotubes, stimuli-responsive vesicles, and helical superstructures were revealed in the solutions. Meanwhile, in recent years, rod molecules have adopted aromatic rod segments because aromatic rods exhibit their particular functions via π–π interaction, in addition more well-defined molecular structures can be obtained by stepwise coupling pathways [36] as compared to non-mesogenic rod segments that mainly rely on polymerization methods [37, 38]. Thus, self-assembling aromatic rod systems would be an ideal model to define correlations between solution properties and molecular parameters. Although non-1-D structures were observed and are interesting, in this review we focus on 1-D assemblies, the so-called nanofibers and related nanostructures, of molecular rods in solutions.

As far as the criterion for sectioning this review article is concerned, it is not straightforward. For convenience, we will review 1-D assemblies of molecular rods in terms of molecular architecture. We classify rod assemblers into six molecular categories; (1) rod–coil molecules, (2) macrocyclic molecules, (3) dendron–rod–coil molecules, (4) dumbbell-shaped molecules, (5) wedge-typed molecules, and (6) conjugate rods with lateral chains. We will describe biomimetic or bioconjugate amphiphilic rods in the last section.

2 One-Dimensional Assemblies of Molecular Rods in Solutions

One-dimensional nanostructures such as fiber, ribbon and tubules in solution states by self-assembly of rod molecules, and their related functions will be introduced.

2.1 Rod–Coil Molecules

Rod–coil molecules are composed of conformationally rigid rod and flexible coil blocks in a molecular structure. Among possible amphiphilic rod candidates, they would have the simplest molecular structure, and thus are easy to approach via simple synthetic pathways. So far, several rod–coil molecules that form one-dimensional fibrous aggregates in solution have been reported.

In comparison to achiral molecules, chiral molecules have optical activities due to their asymmetric chiral centers, resulting in rotating a plane polarized light in a preferred way of either right or left direction. Introduction of an optically active chiral group particularly near the interface of rod/coil building blocks might be able to induce a preferred rod packing direction, leading to a supramolecular chirality, i.e., a helical arrangement of the rod segments. In 2002, Meijer et al. reported on an ABA-typed triblock coil–rod–coil molecule (**1**) that consists of chiral penta(ethylene oxide) coils and a hexathiophene rod [39, 40]. As confirmed by the AFM image of Fig. 1a, the molecules formed chiral helical aggregates in *n*-butanol or water, accompanied by a hypsochromic shift of the UV-Vis absorption, a hypochromic effect of the fluorescence and bisignate CD data.

The helical aggregates show a phase transition at a certain temperature, the so-called "melting temperature" at which it loses the CD activity. However, somewhat different CD behavior between butanol and water solutions was observed. Upon heating, in butanol the aggregates directly transform into

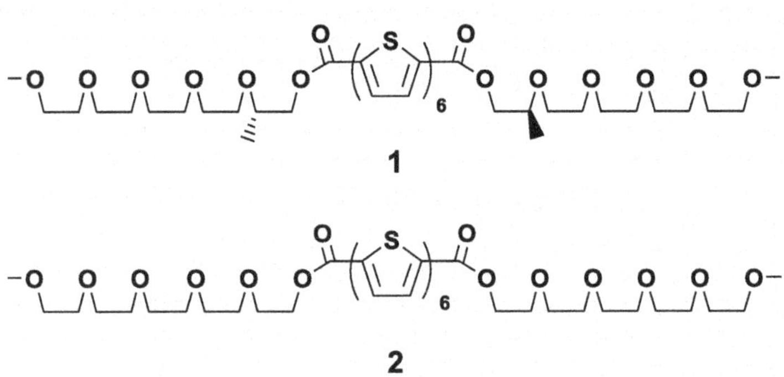

Scheme 1

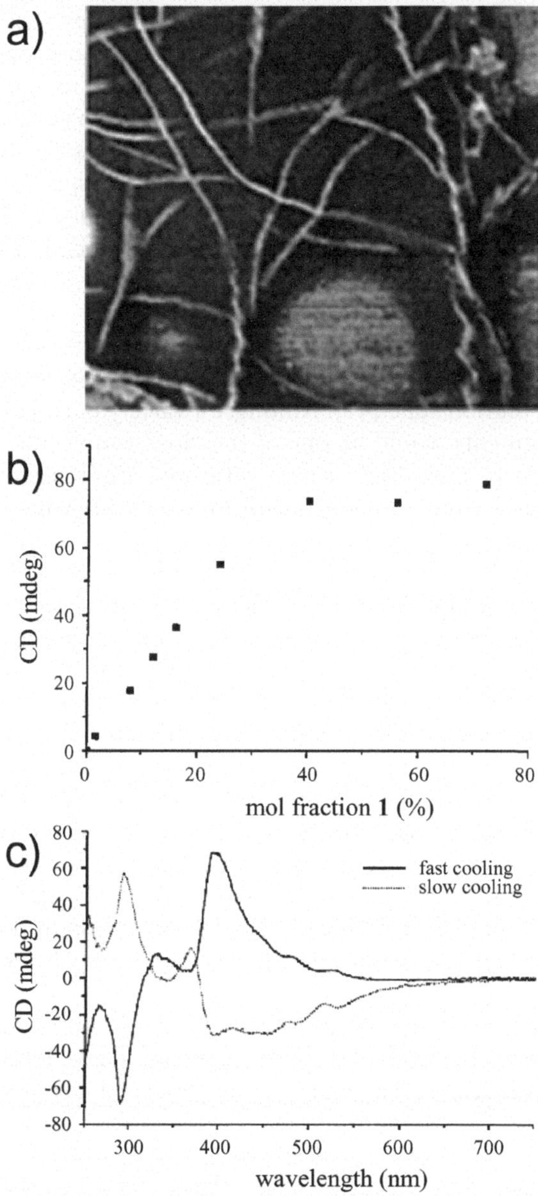

Fig. 1 **a** AFM image of 1, showing left-handed helical aggregates. **b** Cotton effect as a function of the mol of "sergeant" 1 compared to "soldier" 2. **c** Cotton effect observed after fast and slow cooling of a mixture of 1 and 2. Reprinted with permission from [40]. © (2002) American Chemical Society

molecularly dissolved species at 30 °C, while in water they pass through achiral aggregates. In this rod–coil system, Meijer et al. also proved a "sergeant and soldiers" phenomenon which means that chirality in aggregates can be produced in the presence of a small amount of chiral molecules (in this work, compound **1**) in an assembly of major achiral molecules (in this work, compound **2**) (Fig. 1b). On slow cooling, monitoring of CD spectra indicated that a mixture with 25% of **1** shows a similar CD spectrum to that of pure **1** (Fig. 1c). Strikingly, upon fast cooling the chirality inversion was observed, from which the authors proposed that in this case the inversed chirality is favored kinetically.

Lee et al. have employed rod–coil molecular systems consisting of oligo(*para*-phenylene) (OP) rod units for the study of self-assembled nanostructures in the solution state [41]. Amongst a variety of rod building blocks, OP-based rod segments would be one of the closest molecular units to ideal rod segments due to their high aspect ratio and rigid conformation. Lee's group began with a simple coil–rod–coil molecule (**3**) where a linear type hexa-*para*-phenylene (HP) rod is connected to two equivalent hydrophilic poly(ethylene oxide) (PEO) coils at both rod ends. As expected from the endowed molecular amphiphilicity, the molecule was observed to self-assemble into an aggregate structure in the aqueous solution. From the DLS study, the aggregate structure from **3** showed discrete spherical micelles at the early stage, and then slowly transformed into cylindrical fibers over a week. According to TEM analysis, the thermodynamically stable fibers have a uniform diameter of about 10 nm and lengths up to several hundreds of nanometers (Fig. 2a). Meanwhile, the solution containing cylindrical fibers could be gelled by the addition of over 10 wt. % of rod–coil–rod molecule **4**. In this event, the rod–coil–rod molecules are considered to act as supramolecular cross-linkers, the rod segments of which are integrated into aromatic rod domains of cylindrical fibers. Finally, the rod–coil–rod molecules interconnect individual fibers into bundles, as confirmed by the TEM image of Fig. 2b.

Yu et al. reported a rod–coil molecule (**5**) consisting of oligo(phenylene vinylene) rod and PEO coil [42]. The molecule showed remarkable self-assembling features, and self-assembly into cylindrical micelles in THF/water

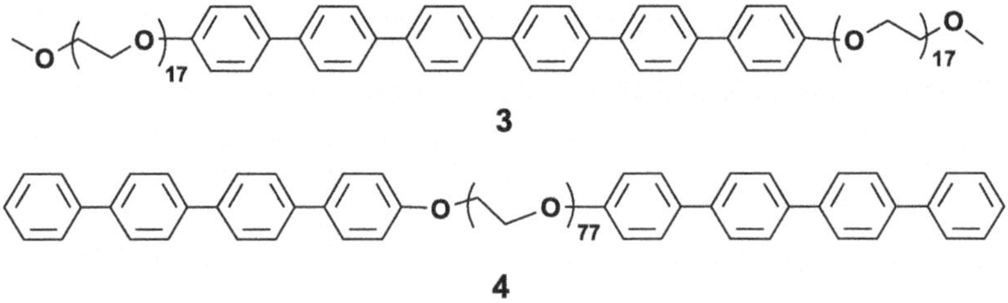

Scheme 2

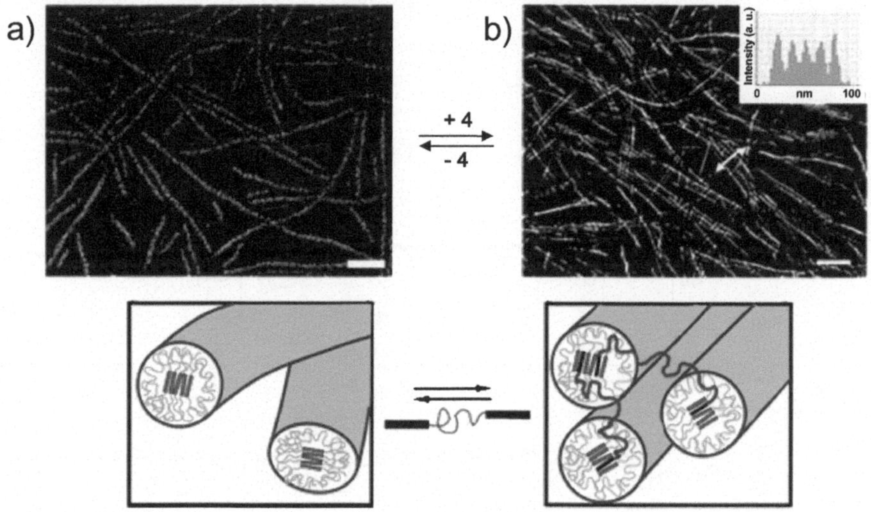

Fig. 2 TEM images of **a** 3 with negative staining (*scale bar* = 50 nm), and **b** nematic gel with 10% **4** relative to **3** (*scale bar* = 100 nm), with density profile *inset*. Schematic representation of reversible bridging between isotropic fluid and nematic gel of **3** with **4**. Reprinted with permission from [41]. © (2005) American Chemical Society

solvents. On the basis of TEM and AFM studies, the cylindrical micelles were observed to be interwoven fibers with lengths up to several thousand nanometers and entangled with one another (Fig. 3a). The OPV cores of the micelles showed a monodisperse diameter of 8–10 nm. In addition, the authors prepared a homologous diblock copolymer (**6**) consisting of the same rod block and poly(propylene oxide) (PPO) coils [43]. Similar to the observed fibrous morphologies from compound **5**, these molecules exhibited interwoven fibers with lengths of several thousand nanometers and diameters of about 8–10 nm. On the basis of SANS, SAXS and TEM results, the authors proposed a packing model in which a hydrophobic OPV inner core is sur-

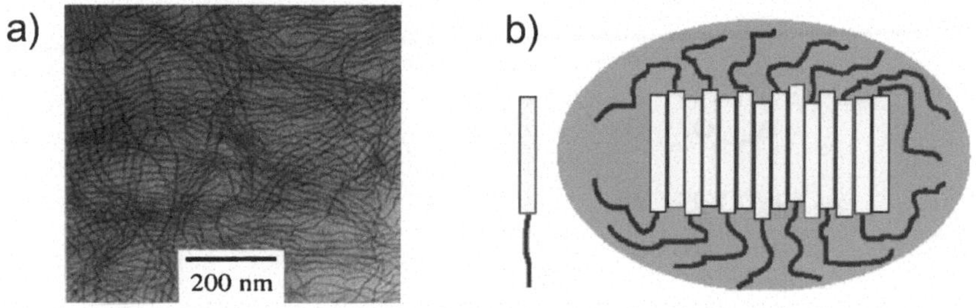

Fig. 3 **a** TEM image of copolymer **5** film. Reprinted with permission from [42]. © (2000) American Chemical Society. **b** Schematic representation of possible packing model for **5–7** copolymers. Reprinted with permission from [43]. © (2004) Wiley

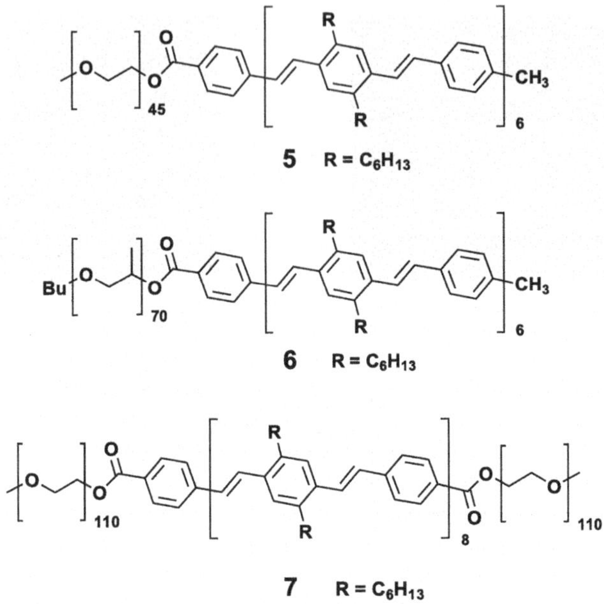

Scheme 3

rounded by a PPO shell. Furthermore, on the basis of the radius of 8.9 nm of the cylindrical fiber assembled by ABA type compound **7**, the authors suggest an interdigitated monolayer packing of **6** (Fig. 3b).

Sung et al. synthesized sugar-based small rod–coil molecules (**8a–d**), and investigated the self-assembly behavior as a function of coil length [44]. Arising from the amphiphilicity between hydrophilic sugar and hydrophobic residual moieties, the chiral rod–coil molecules gave rise to the formation of aggregates in the aqueous solution. As evidenced by TEM and FESEM, an interesting transformation from a platelet (from **8a**) to helical twist morphologies (from **8b–d**) was observed (Fig. 4a,b). According to the authors, the formation of helical twists originated from the bending and twisting forces of the sugar-based chiral entity.

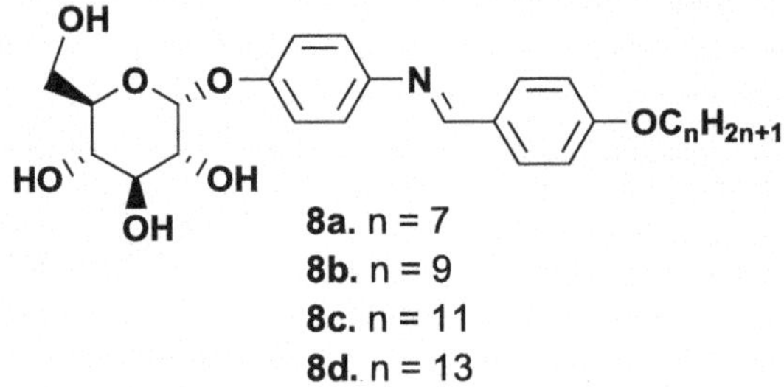

Scheme 4

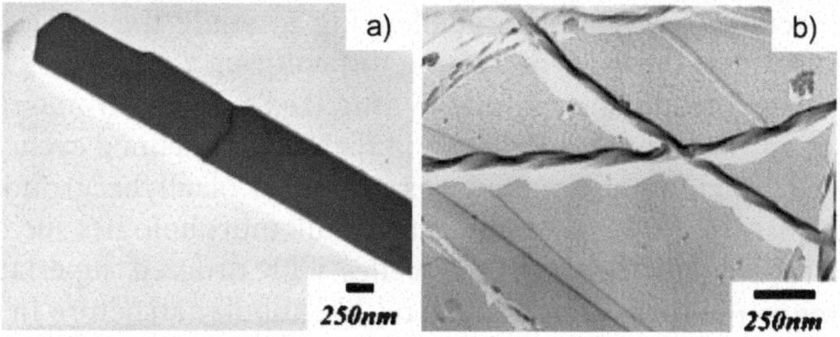

Fig. 4 TEM images of compounds **a** **8a**, and **b** **8d**. Reprinted with permission from [44]. © (2006) American Chemical Society

2.2 Macrocyclic Molecules

As compared to non-cyclic compounds, it is quite probable that cyclic rod–coils show a distinct assembling behavior in bulk and solution, because in a packing structure most coil parts will inevitably be concentrated near rod segments, significantly inhibiting inter-rod packing. Despite such a straight-

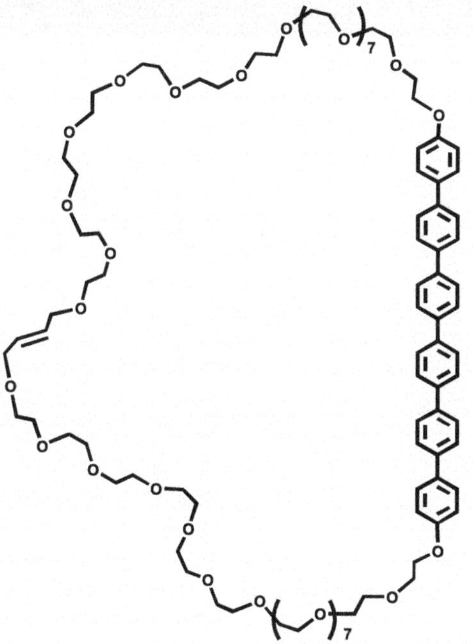

9

Scheme 5

forward expectation, it has been challenging to study the self-assembly of macrocyclic systems due to synthesizing difficulties.

Recently, Lee's group successfully synthesized a rod–coil macrocycle (**9**) consisting of an HP rod and a PEO coil [45]. The well-defined cyclic molecule was prepared by ring-close metathesis reactions of allylic groups bearing rod–coil precursors. Aside from interesting bulk morphologies such as supramolecular ribbons and barrel-like micelles with ordered superlattices, the most striking is the formation of a barrel-like tubular structure in the water environment. In the aqueous solution, the discrete tubular objects have a hydrophilic exterior and interior consisting of PEO coils and water molecules (Fig. 5a,b). On the basis of DLS studies, the barrel-type aggregation showed

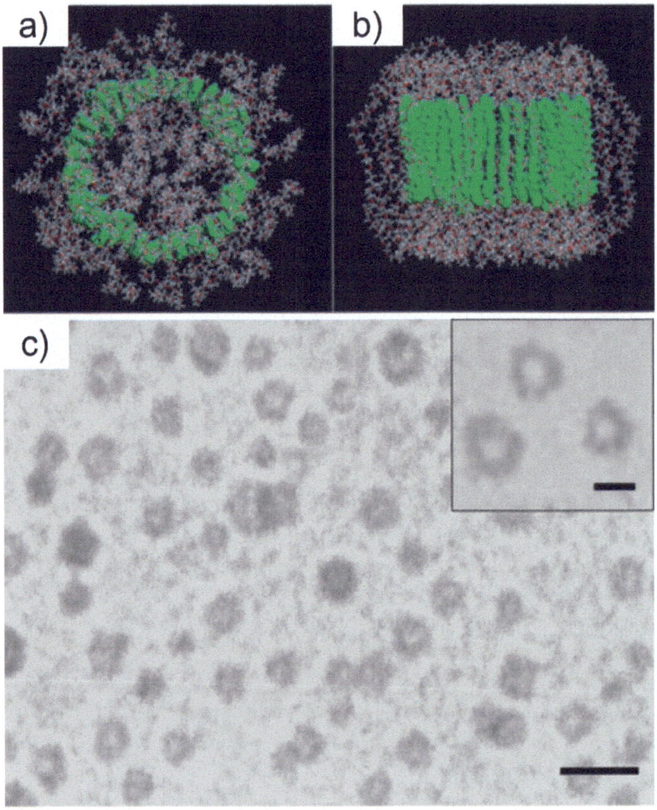

Fig. 5 **a** Top view and **b** side view of a barrel-like structure by molecular modeling of **9** (rod segments are colored *green*, oxygen and hydrocarbons are colored *red* and *grey*, respectively). **c** TEM image of barrel-like tubular structures of **9** in an aqueous solution. The *scale bar* represents 20 nm. *Inset*: higher-magnification image showing contrast between the periphery and center. The *dark areas* arise because the rod segments are stained more than the coils. The *scale bar* represents 5 nm. Reprinted with permission from [45]. © (2005) Nature Publishing Group

a uniform size of average diameter of about 7 nm. From TEM observations of a distinctive contrast between the periphery and center in the object, a barrel structure with a uniform hydrophilic channel was suggested for the aqueous solution (Fig. 5c).

As inspired by helical and tubular architectures observed in proteins and nucleic acids, Lee et al. prepared a modified macrocycle (**10**) which has chiral centers near both termini of the HP rod [46]. The preparation of the chiral macrocycle was implemented like the achiral macrocycles except for employing chiral PEO coils. The authors investigated supramolecular assembly behavior dependent upon chirality introduction, using circular dichroism (CD) spectra in the aqueous solution.

CD data above a certain concentration (from 0.005 wt. %) showed a negative Cotton effect first, followed by a positive Cotton effect at higher wavelength with the CD signal passing through zero near the absorption max-

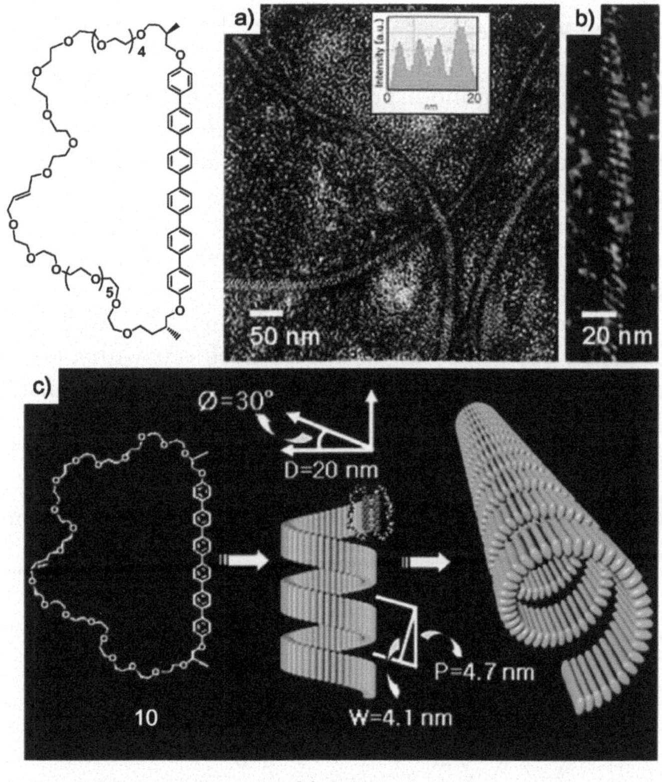

Fig. 6 TEM images of **a** negatively stained left-helical tubular structure of **10** with density profile *inset*, and **b** magnification of the left-helical tubular structure. **c** Schematic representation of a proposed mechanism for the formation of the helical tubular structure of **10**. Reprinted with permission from [46]. © (2006) American Chemical Society

imum of the chromophore, suggesting the formation of a helical superstructure with a preferred handedness. TEM analysis of the unstained sample demonstrated fibrous aggregates with lengths up to several micrometers and a uniform diameter of about 20 nm (Fig. 6a). Through more detailed interpretation of individual fibers, they were shown to feature helical tubules (Fig. 6b). The internal diameter and the wall thickness were estimated to be 14 and 3 nm, respectively. Notably, TEM images of the samples stained with uranyl acetate showed tubules with a left-handed helical arrangement and a regular pitch (4.7 nm) corresponding to the extended molecular length (Fig. 6c). The authors explained the observed tubular structure with coiled ribbons by interplay of the energy balance between repulsive interactions among the adjacent flexible chains and π–π stacking interactions. The mechanism suggested is that macrocycles self-assemble in a coiled fashion to form a helical tubular structure, relieving space crowding of coil segments without sacrificing a parallel arrangement of the rod segments.

2.3 Dendron–Rod–Coil Molecules

Dendritic blocks with a monodisperse branched shape would be very versatile building units for the engineering of supramolecular architectures and functions because of their molecularly uniform fanwise shape and the ability to introduce multiple functional groups [47].

Stupp et al. reported a so-called dendron–rod–coil (DRC) (**11**) by introduction of an aromatic dendron moiety with hydroxyl peripheries into a diblock rod–coil molecule [48, 49]. The DRC molecules formed a one-dimensional aggregate structure in the CH_2Cl_2 solution. When cast from a 0.004 wt. % CH_2Cl_2 solution, a TEM analysis observed 1-D objects with a uniform width of 10 nm and lengths on the order of micrometers (Fig. 7a). Considering that the fully extended length of a DRC molecule is about 6.5 nm, the estimated width suggests a dendron head-to-dendron head bimolecular packing of DRC molecules (Fig. 7b). This is consistent with the crystal structure of the model compound where eight hydrogen bonds are formed into a tetrameric cross-

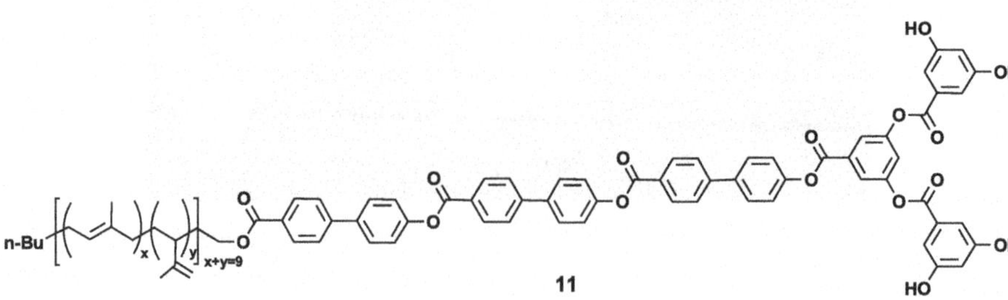

11

Scheme 6

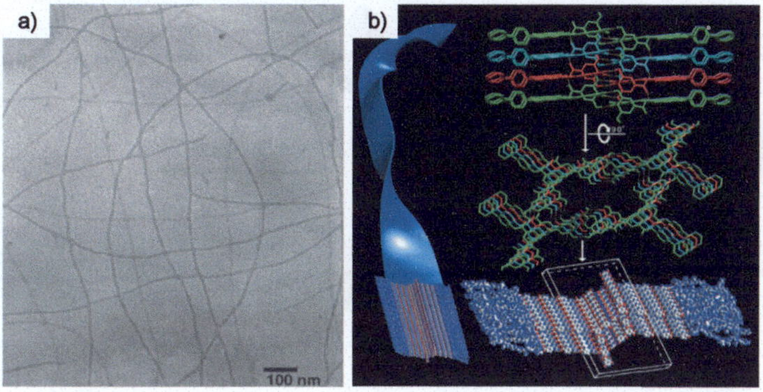

Fig. 7 **a** TEM image of unstained **11** nanoribbons formed in dichloromethane. **b** Schematic representation of the proposed structure for **11** nanoribbon. Reprinted with permission from [48]. © (2001) American Chemical Society

section of the ribbon [50]. The AFM image of the 1-D object revealed the thickness of 2 nm, indicative of a ribbon-type fiber. These DRC molecules were observed to undergo gelation, producing a soft solid with a blue-violet hue in styrene, 2-ethylhexyl methacrylate (EHMA) as well as CH_2Cl_2 [51]. The polymerizable styrene solution containing 1 wt. % of DRC molecules was heated to polymerize the monomer. In marked contrast to ordinary polystyrenes, in the gel state the polymerized solid revealed signatures of order such as birefringence and fiber formation with much greater molecular orientation, indicating that a small quantity of supramolecular nanoribbons function as a structural scaffold in the media (Fig. 8). To investigate the role of each block for the gelation, the authors altered a number of hydroxyl groups and biphenyl units, and coil components. By comparing the gelation

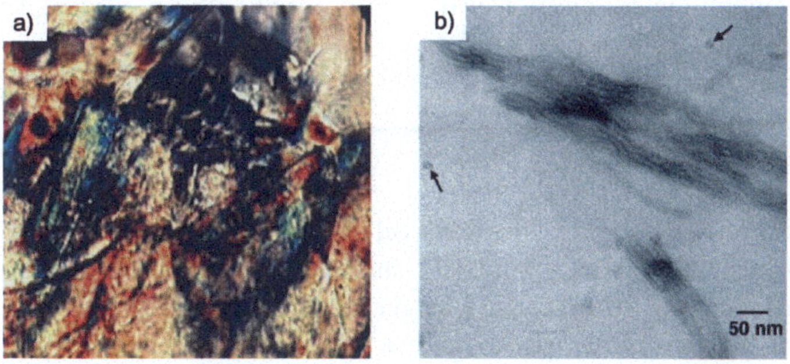

Fig. 8 **a** An optical micrograph between cross-polarizers shows the scaffolded solid polystyrene has a birefringent texture (1 wt. % of **11**) at 100 °C. **b** TEM image of a thin film (75 nm thick) ultramicrotomed from the scaffolded material (containing 1 wt. % **11**) and stained with osmium tetroxide. Reprinted with permission from [51]. © (2002) Wiley

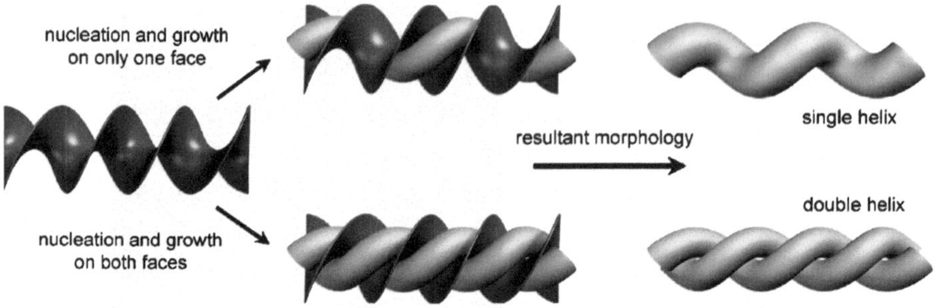

Fig. 9 Schematic representation of templating pathways. Nucleation and growth on one side of the twisted ribbons (*dark*) leads to single helices of CdS (*bright*), while nucleation and growth on both sides of the ribbon leads to double helices. Reprinted with permission from [53]. © (2005) Wiley

tendencies of respective molecules, they demonstrated that gelation of DRC molecules occur by interplay of π–π interaction between aromatic rods and directional hydrogen bonding.

Stupp et al. investigated the details of supramolecular assembled structures upon solvent variation. In contrast to the flat nanoribbons in the CH_2Cl_2 solution, they found that nanoribbons in other solvents such as ethyl methacrylate (EMA) or 2-ethylhexyl methacrylate (EHMA) twist into helical ribbons. The authors took advantage of hydrophilic core regions composed of dendron hydrogen-bonding sites for the mineralization of CdS [52, 53]. As evidenced by TEM and ED studies, grown CdS nanocrystallites templated by twisted helical ribbons turned out to be either single- or double-stranded coiled-helix with a pitch of 40–60 nm. The authors suggested that the twisted helical ribbons act as structure-directing agents for the formation of CdS with well-defined nanoarchitectures due to the affinity of Cd^{2+} ions for the hydrophilic dendron region in hydrophobic environments (Fig. 9). Consequently, the intact twisted helical shape of the ribbons could be interpreted to govern the peculiar coiled helical morphology of CdS by guiding the crystal growth along the face of the template.

2.4
Dumbbell-Shaped Molecules

As an alternative approach, flexible coil parts in rod–coil molecules can be structurally modified into dendritic shapes. In comparison to linear-type coils, dendritic coils would cause a larger steric repulsion at the rod/coil interface, which influences rod-packing structures. Lee et al. devised a dumbbell-shaped molecule (**12**) based on an elongated dodeca-*para*-phenylene rod block and aliphatic polyether dendritic coils with chiral carbon centers, and investigated the self-assembly behavior in aqueous solution [54]. In remarkable contrast to ordinary one-dimensional fibers, the aggregate structures

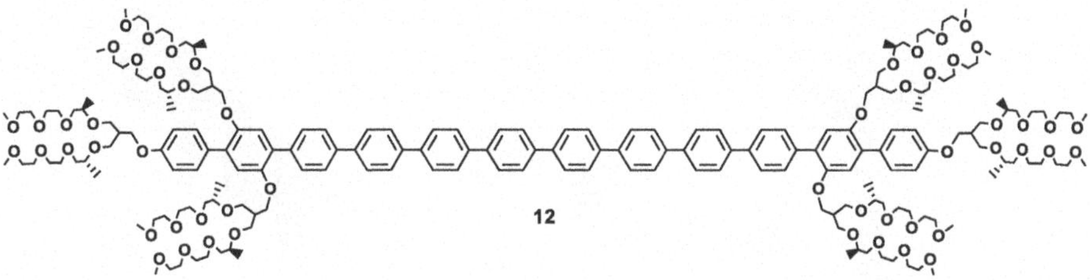

Scheme 7

from the dumbbell molecules exhibited helical cylinders with left-handedness (Fig. 10a). As confirmed by TEM and CD data, the helical fibers have a diameter of a molecular length, a pitch of 5.6 nm and lengths up to several micrometers (Fig. 10b). The authors proposed that the helical fibers are driven by a microphase separation between the incompatible molecular components and directional rod packing associated with the bulky chiral dendritic coils.

Moreover, supramolecular assemblies of another dumbbell molecule (**13**) were investigated by Lee et al. [55]. Similar to the above dumbbell (**12**), these dumbbells showed helical fibers in the aqueous solution. However, in this study the authors emphasized that the helical fibers turn into nanocages with the addition of aromatic guest molecules such as 4-bromonitrobenzene which are selectively intercalated between the rod segments. From a thermo-

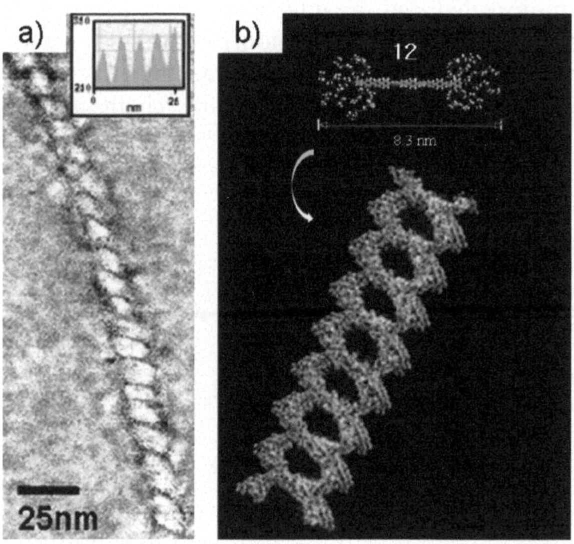

Fig. 10 **a** TEM image of **12** with negative staining, with density profile *inset*. **b** Schematic representation of a proposed mechanism for the formation of the helical nanofiber of **12**. Reprinted with permission from [54]. © (2005) American Chemical Society

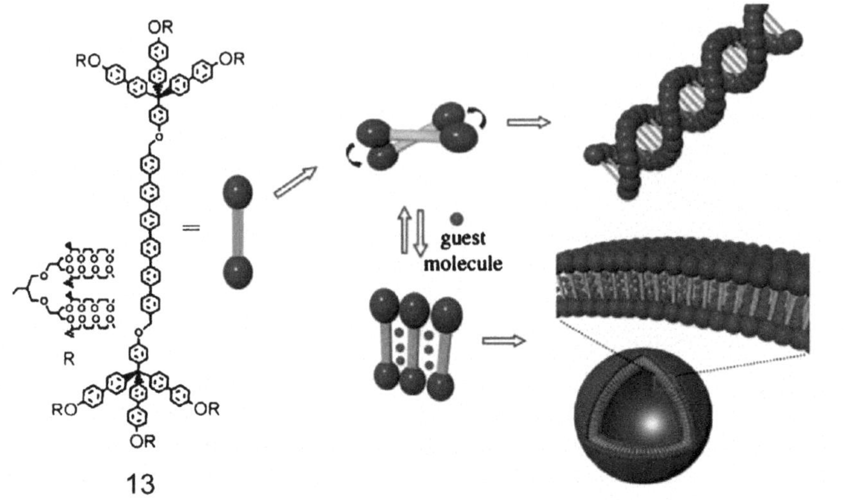

Fig. 11 Representation of the reversible transformation of helical fiber into a spherical capsule. Reprinted with permission from [55]. © (2006) Wiley

dynamic point of view, the intercalated guest molecules would leave some distance between the vicinal dumbbells, relieving the steric hindrance between bulky dendritic segments. The authors proposed that such a packing variation drives the twisted helical arrangement into a parallel arrangement, leading to a transformation into a cage-like structure (Fig. 11). Interestingly, the authors demonstrated that the inter-conversion between the helical fiber and the nanocage is feasible upon addition/removal of the guest molecules.

Unlike symmetric dumbbells, ABC-type asymmetric dumbbells (**14a–c**) with hydrophilic (A) and hydrophobic (C) bulky chains at the rod ends were

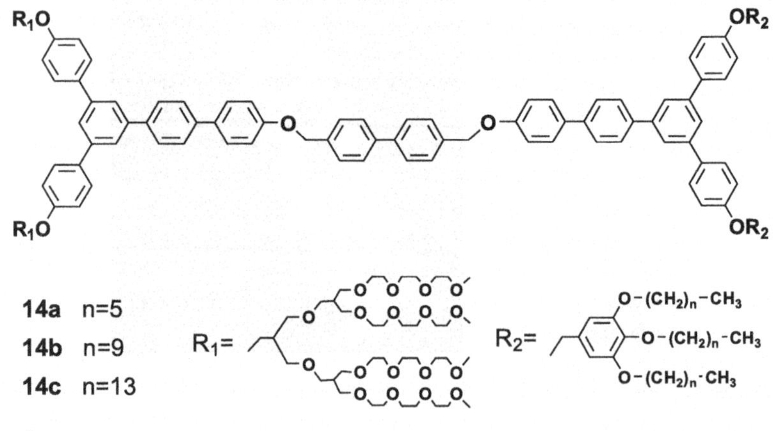

Scheme 8

also prepared by Lee and his co-workers [56]. The authors examined the self-assembly behavior of these asymmetric dumbbells in 0.01 wt. % aqueous solution as a function of hydrophobic alkyl chain length. From TEM studies with cast samples stained with uranyl acetate, compound **14a** with the shortest hexyl hydrophobic chains demonstrated the coexistence of spherical and short cylindrical micelles with diameters of about 15 nm, indicative of interdigitated bilayer packing of molecular dumbbells (Fig. 12a). Compound **14c**, with the longest tetradecyl chains, displayed elongated and entangled cylindrical micelles with a uniform diameter of 18 nm and lengths of at least several micrometers (Fig. 12c). More remarkable phenomena were found in compound **14b** with intermediate decyl chains. The molecules in the same concentrated aqueous solution formed unprecedented nanorings which have a cross-sectional diameter of 16 nm, and ring diameters of 70 to 300 nm

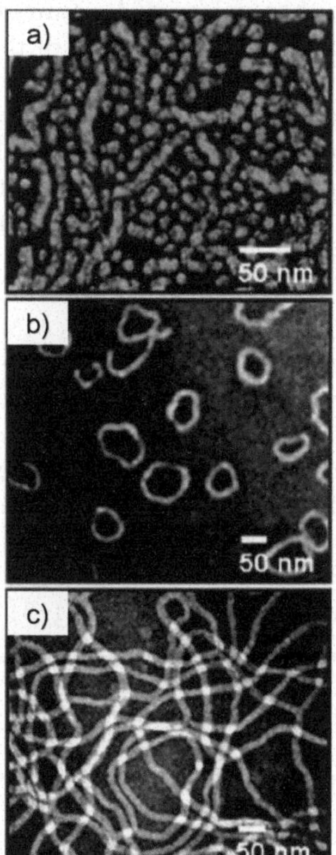

Fig. 12 TEM images of **a** spherical, short cylindrical morphology of **14a** and **b** toroidal morphology of **14b** in aqueous solutions (0.01 wt. %), **c** long cylindrical morphology of **14c** in an aqueous solution (0.01 wt. %). Reprinted with permission from [56]. © (2006) American Chemical Society

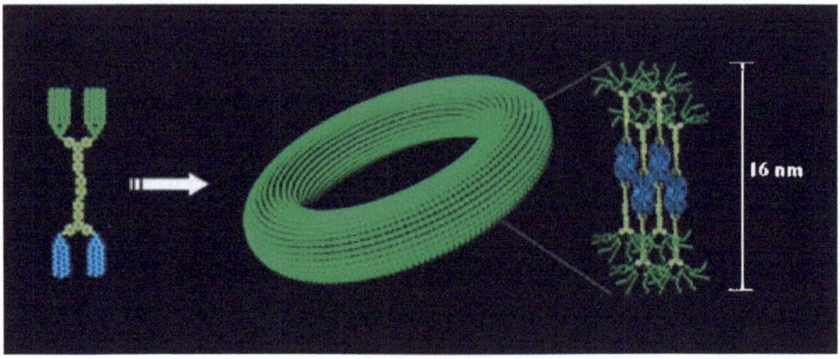

Fig. 13 Schematic representation of nanorings from amphiphilic molecular dumbbell **14b**. Reprinted with permission from [56]. © (2006) American Chemical Society

(Figs. 12b and 13). The author performed a time-dependent TEM in order to gain insight into the formation mechanism of these unusual toroidal aggregates. Consequently, they found that kinetically driven spherical micelles at the initial stage spontaneously transform into thermodynamically stable toroids for a week. The authors suggested that the hydrophobic effect associated with hydrophobic alkyl chain length dominates the overall interfacial area, thus giving rise to the transformation of discrete spherical, toroidal into elongated cylindrical aggregates.

Very recently, Würthner et al. reported a series of rigid perylene bisimide-based wedge- and dumbbell-shaped molecules (**15** and **16a,b**) [57]. All the molecules showed aggregate structures in THF–water mixture solutions, confirmed by excimer formations at 600–800 nm in the fluorescence spectra. Detailed morphological studies through TEM experiments found that the AB-type wedge molecule **15** shows spherical micelles with a diameter of 4–6 nm, which are correlated with Israelachvili's packing parameter of 0.252 for **15** (Fig. 14a) [58]. Symmetric dumbbell molecules **16b** exhibited one-dimensional fibers due to the non-conical molecular shape (Fig. 14c). Interestingly, the authors observed that a mixture (**15** : **16a** = 8 : 1, molar ratio)

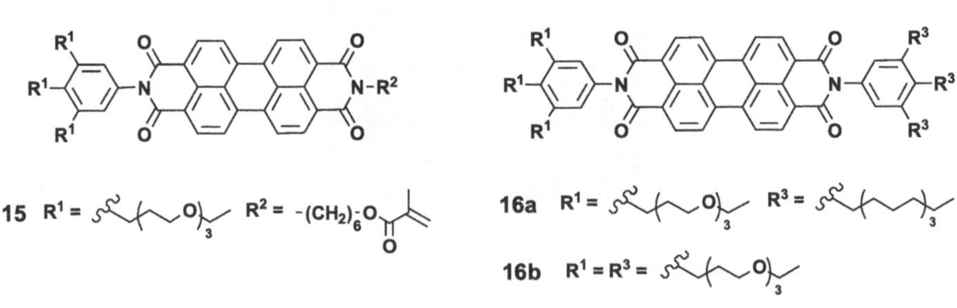

Scheme 9

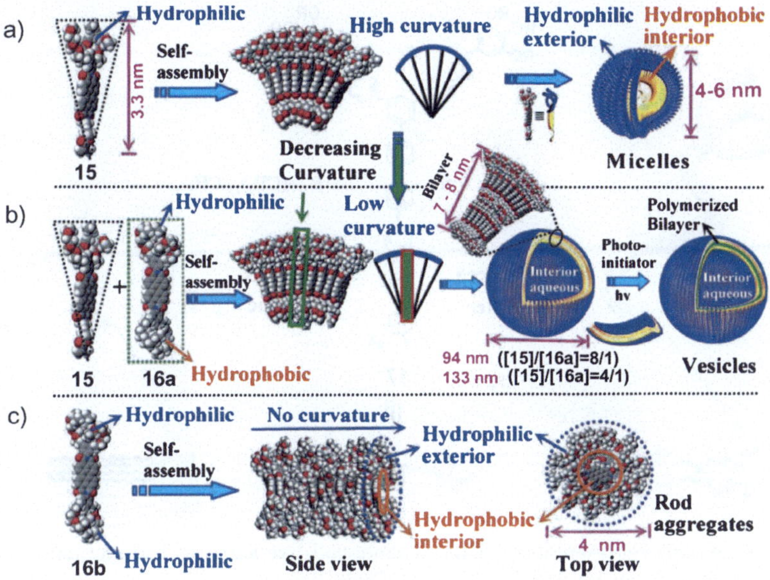

Fig. 14 Schematic illustration for the formation of micelles from wedge-shaped **15** (*top*), bilayer vesicles from the co-self-assembly of **15** and dumbbell-shaped **16a** (*middle*), and rod aggregates from dumbbell-shaped **16b** (*bottom*). Reprinted with permission from [57]. © (2007) American Chemical Society

of molecular wedge **15** and asymmetric ABC-type dumbbell **16a** co-assemble into hollow vesicles with the average diameter 94 nm and a wall thickness of 7–8 nm, and an increase of **16a** relative to **15** (**15** : **16a** = 4 : 1, molar ratio) leads to the expansion of vesicle diameter up to 133 nm (Fig. 14b). The authors explained the observed co-assembly behavior in terms of the change of the interfacial curvature depending on the molar ratio of the wedge and the dumbbell molecules.

2.5 Wedge-Type Molecules

Analogous to dendritic-rod molecules, Lee et al. designed a tree-like wedge-coil (**17**) with a wedge-shaped rigid aromatic segment [59]. Interestingly, they proved that the stiffness of nanofibers assembled by the wedge-coil molecules can be switched as a function of solvent polarity. From TEM and DLS studies, the molecules showed one-dimensional fibrous assemblies in water and hexane solutions (0.01 wt. % in both) because of the amphiphilic characteristics. Further in-depth analysis of the TEM images indicated that the fibers in the aqueous solution look like random coils, while the fibers in the hexane solution form rigid rods with a preferred orientation (Fig. 15). On the basis of DSC and WAXS results, the authors stated that the stiffness contrast

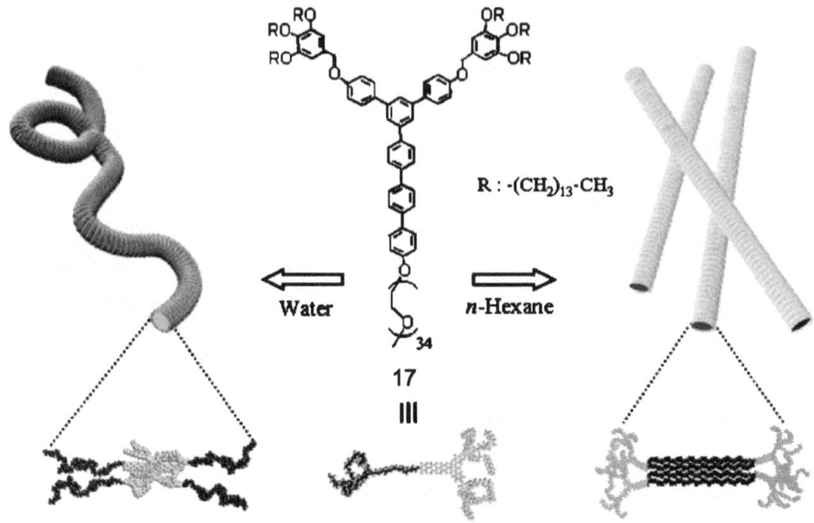

Fig. 15 Schematic representation of the flexible coil-like and stiff rod-like nanofibers of 17. Reprinted with permission from [59]. © (2006) Wiley

that depends on solvent polarity is attributed to the crystallization feasibility of components in the cylindrical core regions of respective supramolecular fibers.

Wedge-type molecules (**18a,b**) bearing a diamino triazine group with multiple hydrogen bonding sites were prepared by Meijer et al. [60]. These molecules were found to form hydrogen-bonding-mediated chiral hexameric rosette structures at the solid–liquid interface using graphite as a substrate and 1-phenyloctane as a solvent. Chiralities of the rosettes, i.e. rotation directions, could be tuneable depending on the molecular chirality, as confirmed by STM images (Fig. 16a,b). The authors speculated that the rosette structure observed on the substrate can be extended to tubular fibers by stacking themselves via π–π interaction in solutions. As characterized by absorption, AFM, CD and SANS data, they found that fibrous aggregates in hydropho-

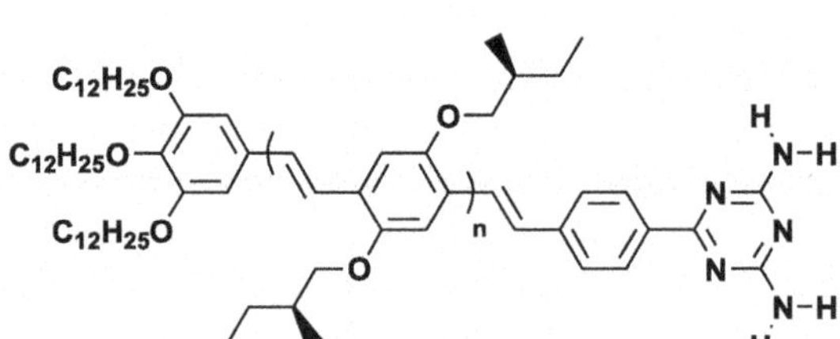

18a n = 1
18b n = 2

Scheme 10

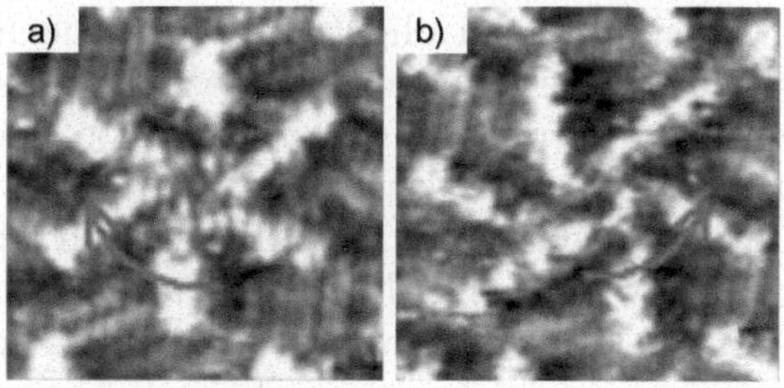

Fig. 16 STM images of **a 18a** and **b 18b** rosettes. *Red arrows* indicate the rotation direction of the rosette. Reprinted with permission from [60]. © (2004) Wiley

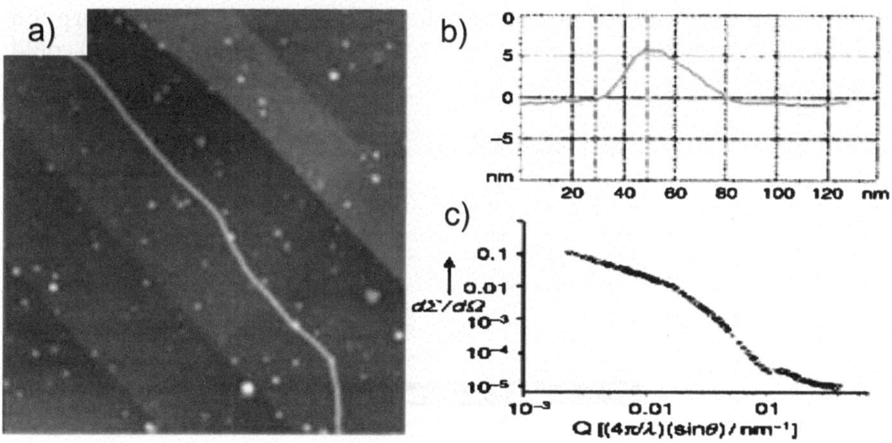

Fig. 17 **a** AFM image of drop-cast **18b** solution in heptane on graphite, **b** height profile of a fibril, and **c** SANS data revealing columnar structures in dodecane solutions of **18b**. Reprinted with permission from [60]. © (2006) Wiley

bic solvents (heptane and dodecane) are optically chiral, and the diameter of the fibers (~6.4 nm) is consistent with the diameter of the hexameric rosettes from the STM image (Fig. 17). From these experimental observations, the authors suggest the formation of self-assembled tubular aggregates in the hydrophobic solvents.

2.6 Conjugate Rods with Lateral Chains

Modified conjugated rod segments with lateral chains can be rod building blocks that self-assemble into a variety of supramolecular one-dimensional fibers and related structures. Furthermore, if rod segments are composed of electro-optically active conjugate moieties, the self-assembled nanostructures would provide interesting supramolecular platforms for versatile functional materials [12, 61, 62].

The Ajayaghosh group adopted oligo(*para*-phenylenevinylene)s (OPVs) (**19a,b**) as self-assemblers where apolar alkyl chains are attached as lateral groups, and polar hydroxyl groups exist at both ends of the rod [63]. From absorption and fluorescence spectroscopic data, Ajayaghosh et al. proved that amphiphilic OPV **19a** form fibrous gel nanostructures in a hydrophobic solvent such as hexane. The SEM image showed entangled nanostructures with a diameter of 100–150 nm (Fig. 18a), and XRD exhibited d-spacings of 23.4 Å in the small-angle region, and 3.8 Å in the wide angle-region, indicating a π-stacked lamellar packing. On the experimental evidence, the authors concluded that cooperative hydrogen bonding and π–π stacking forces are crucial factors for the formation of gel nanostructures (Fig. 18b). The authors took advantage of the gels to devise light-harvesting materials [64]. Particularly, absorption and emission studies demonstrated that OPV **19b** and Rhodamine B (**20**) could

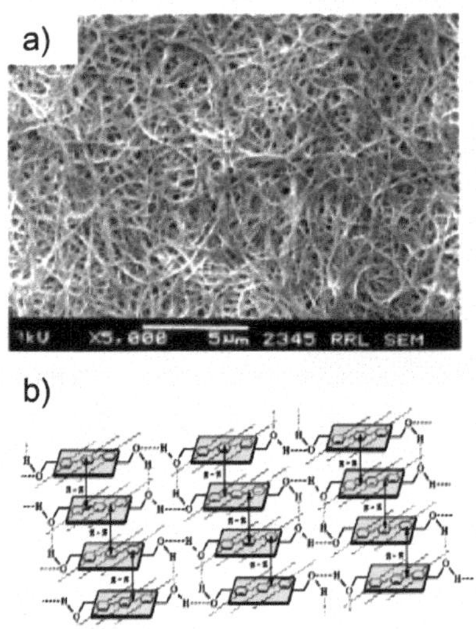

Fig. 18 **a** SEM image of a dried gel of **19a** from toluene. **b** Probable self-assembly of the gel in hexane. Reprinted with permission from [63]. © (2001) American Chemical Society

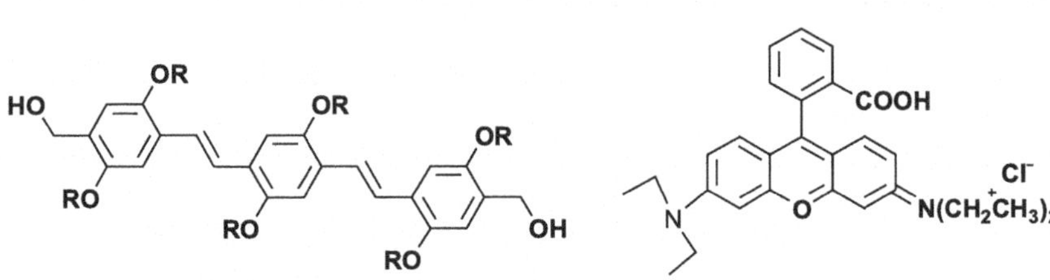

19a: R = $C_{12}H_{25}$
19b: R = $C_{16}H_{33}$

20

Scheme 11

be ideal energy donor and acceptor molecules for a light-harvesting system, respectively. The authors prepared a light-harvesting gel material with the addition of rhodamine B into the OPV gel. Surprisingly, an enhancement of energy transfer in the gel state was observed, which was explained by the fact that the gel-trapped dye molecules are within the Förster radii (Fig. 19).

Also, the group prepared energy donor (**21**) and acceptor (**22**) rod molecules which can form aggregates in decane, and by complexation showed an efficient fluorescence resonance energy transfer (FRET) system [65]. Co-assembly of the donor and acceptor molecules produced stable organogels. Interestingly, these gels radiated different colors, from green to red, depending on the amount of added acceptor molecules (0–20 mol %) (Fig. 20a). The authors described that first, a fast exciton migration from donors to acceptors occurs, and statistically distributed aggregates of acceptor **22** with a differ-

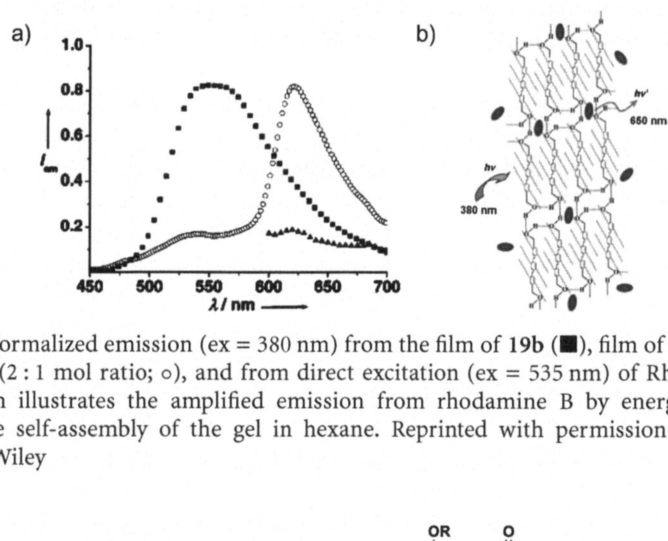

Fig. 19 **a** Normalized emission (ex = 380 nm) from the film of **19b** (■), film of **19b** + Rhodamine B (2 : 1 mol ratio; ○), and from direct excitation (ex = 535 nm) of Rhodamine B (▲), which illustrates the amplified emission from rhodamine B by energy transfer. **b** Probable self-assembly of the gel in hexane. Reprinted with permission from [64]. © (2003) Wiley

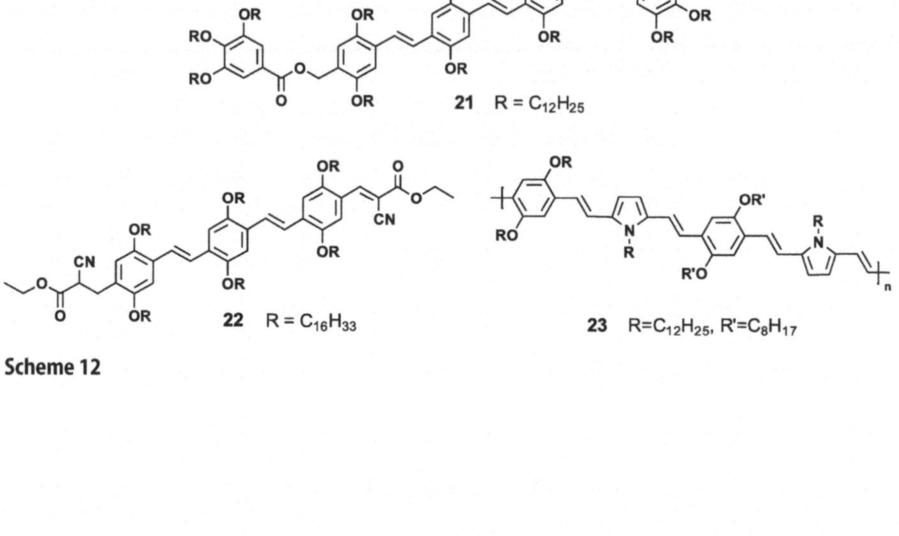

Scheme 12

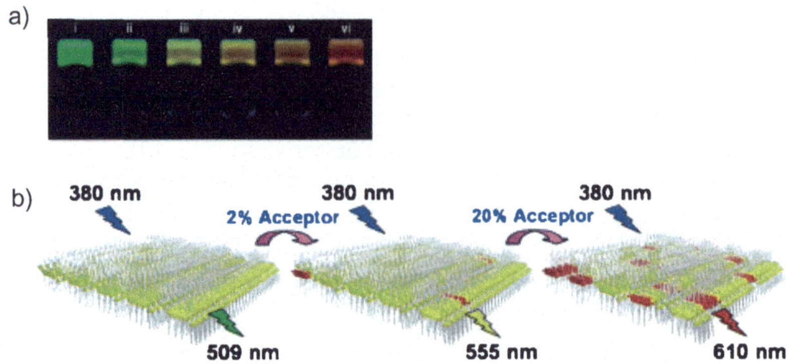

Fig. 20 **a** Photographs of gels of **21** and **22** at different compositions of **22** (i) 0 mol %, (ii) 1 mol %, (iii) 2 mol %, (iv) 6 mol %, (v) 12 mol %, and (vi) 20 mol % under illumination at 365 nm. **b** Schematic representation of the FRET process within the coassemblies having different amounts of the acceptor. Reprinted with permission from [65]. © (2006) American Chemical Society

ent HOMO–LUMO gap produce an energy gradient (Fig. 20b). As a further extension, the group utilized a conjugate polymer (**23**), so-called PYPV, as an acceptor instead of mini-rod molecules for an efficient supramolecular light-harvesting antenna [65]. From this approach, the authors observed a remarkable FRET efficiency by encapsulating an *extremely* small amount of **23** (1.53 mol %) within a donor gel scaffold. The remarkable overlap between the emission band of the donor and the absorption band of the acceptor as well as the large extinction coefficient of the acceptor, satisfied the conditions for a high-efficiency FRET system. OPV derivative **19b** with hexadecyl chains as energy donors formed strong gels in cyclohexane which were maintained after adding a small amount of **23**, as evidenced by observed ribbon-type fibers in fluorescence microscopy and AFM images (Fig. 21a,b). By combination of the OPV fluorescence decay profiles where the emission decay of **19b** was observed to be faster with increasing **23**, and the absorption/emission spectra, the authors concluded that the designed gel-assisted FRET system could be an efficient artificial light-harvesting antenna. Also, they emphasized that the result is the first example employing a semiconducting molecular wire as an energy trap (Fig. 21c).

To develop supramolecular chiral assemblies, Ajayaghosh et al. simply modified the molecular structure by addition of chiral groups at the lateral chains [67]. Although the presence of chiral groups generally reduces the possibility of gel formation, molecule **24** with two chiral groups at the middle phenyl unit formed gels consisting of helical fibers. The authors found an interesting hierarchical assembly of helical fibers, monitored by CD signals, as a function of either concentration or temperature. At low concentrations or higher temperatures, the molecules formed randomly distributed helical aggregates as suggested by weak non-bisignate CD signals. At higher con-

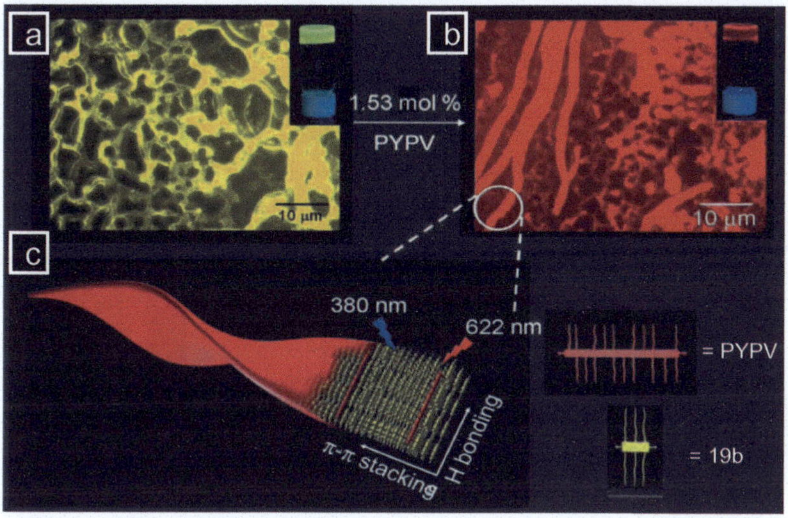

Fig. 21 Fluorescence microscopy images of the drop-cast **19b**-cyclohexane gel (1.12 mM): **a** in the absence and **b** presence of **23** (1.53 mol %), the *insets* show photographs of the gels under the respective conditions when illuminated at 365 nm. **c** A schematic representation of a **23**-encapsulated **19b** tape. Reprinted with permission from [66]. © (2007) Wiley

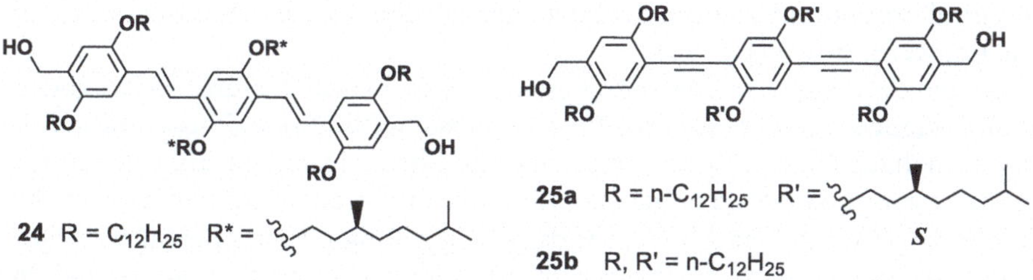

Scheme 13

centrations or low temperatures, an observed strong bisignate Cotton effect indicated a well-ordered helical architecture, which was further confirmed by an AFM image revealing coiled-coil rope-like structures of individual fibers (Fig. 22). Ajayaghosh et al. also prepared chiral and achiral molecules (**25a,b**) composed of an oligo(*para*-phenyleneethynylene) (OPE) rather than OPV [68]. The authors demonstrated a particular reversible supramolecular transition from vesicles to helical tubules in decane via a "sergeant-and-soldiers" coassembly approach. According to CD data, the respective solutions of the achiral and the chiral compounds did not show any Cotton effect, while mixture solutions (1×10^{-5} M) containing 5–30 mol % of chiral compound **25a** resulted in a positive Cotton signal and two negative signals. Structural details of the chiral aggregate were visualized by AFM and TEM

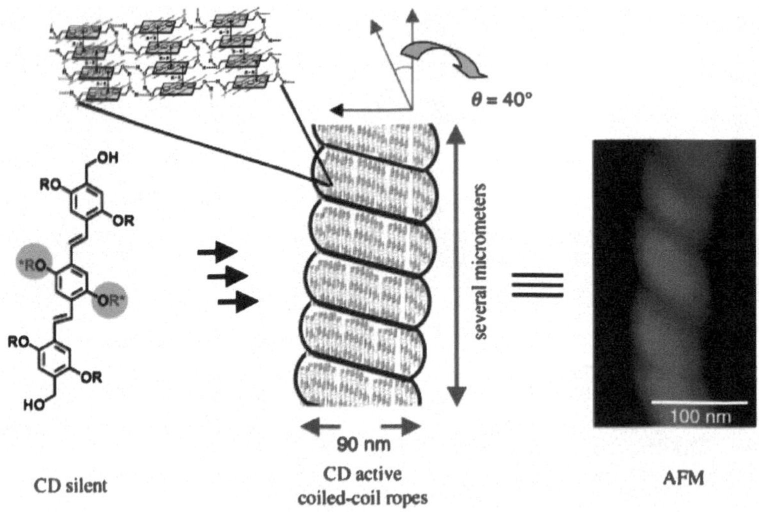

Fig. 22 Schematic representation of the hierarchical self-assembly of **24** into helical coiled-coil gel nanostructures. A magnified AFM image of the coiled-coil rope is shown on the *right*. Reprinted with permission from [67]. © (2004) Wiley

studies. Vesicular aggregates with an average diameter of 94 nm were observed in the achiral compound solution, while in the solution with 25 mol % of chiral compound **25a** only tubular helical fibers with a uniform pitch of 140 nm were shown (Fig. 23).

As another approach to the engineering of helical features, Ajayaghosh et al. designed monocholesterol-functionalized OPV (**26**) and bis-cholesterol-functionalized OPV (**27**) molecules, and then investigated packing structures in the organogels [69]. The two compounds showed distinct behavior in the gelation, CD spectra and optical texture, suggesting a fundamental morphological difference between the two. AFM studies showed that compound **26**

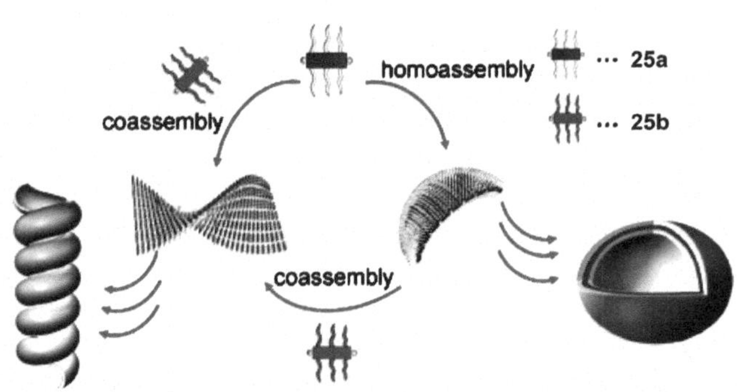

Fig. 23 Self-assembly of vesicles and subsequent transformation into helical tubes in decane. Reprinted with permission from [68]. © (2006) Wiley

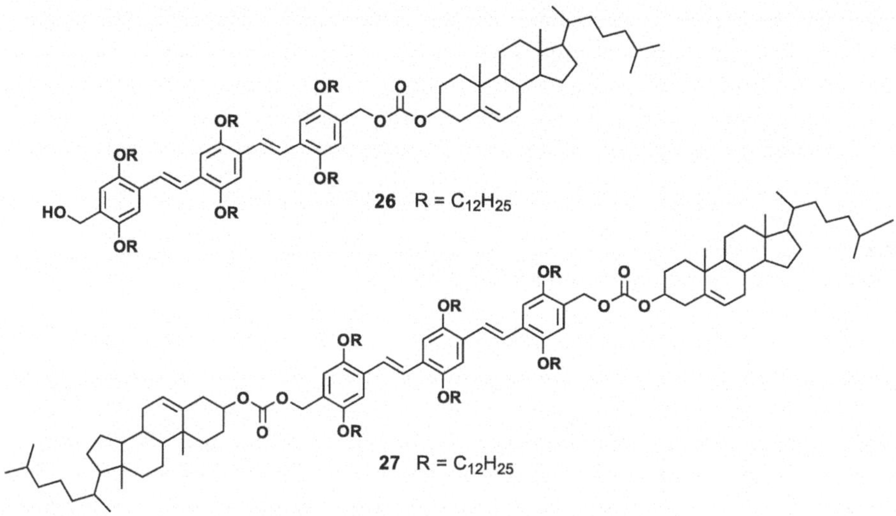

Scheme 14

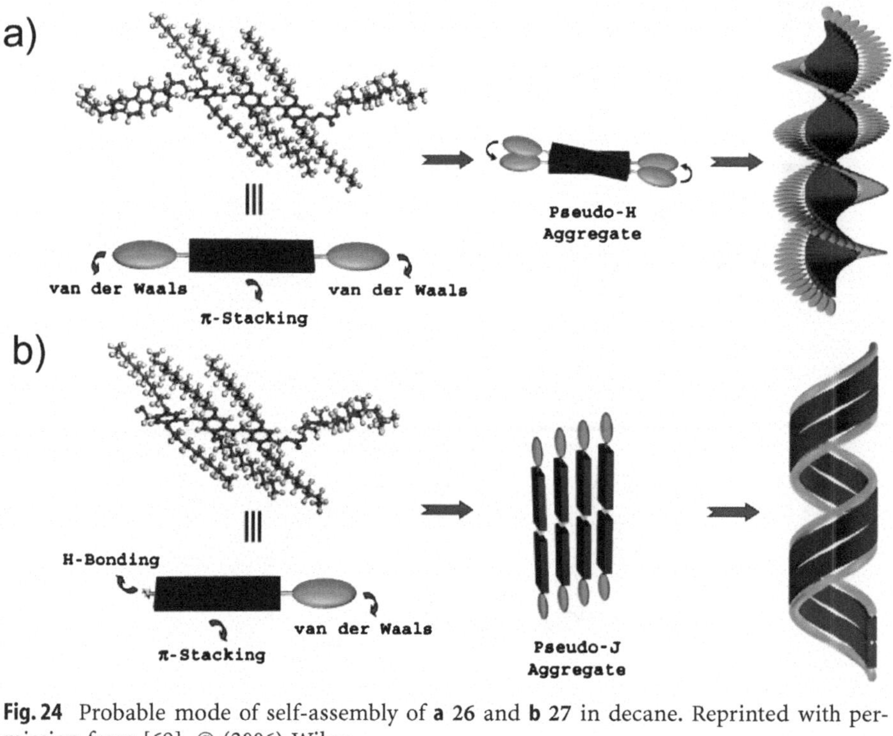

Fig. 24 Probable mode of self-assembly of **a** **26** and **b** **27** in decane. Reprinted with permission from [69]. © (2006) Wiley

organizes into ribbon-like fibers; on the other hand, compound **27** assembles into branched and entangled helical fibers. On the basis of CD, emission, absorption and AFM data as well as molecular architectural consideration, the author suggested that a hydrogen-bonding-assisted coiled helix is formed by a pseudo-J aggregation of **26**, and **27**, with two bulky cholesterol groups, organizes into a twisted helix in a pseudo-H aggregation (Fig. 24).

2.7
Biomimetic or Bioconjugate Amphiphilic Rods

Self-assembly phenomena based upon secondary forces have been widely observed in biological systems such as DNA, viruses, bacterial cell surface layers etc [70–73]. Nature-made supramolecular objects are very sophisticated in size and shape, and show tremendous specific and efficient functions. For example, we know many enzymes are responsible for specific catalytic processes. In this context, creation of supramolecular architectures with biomimetic or bioconjugate features would be desirable and worth investigating. In particular, a few examples related to the title of this review article have been reported.

Recently, Shimizu et al. prepared a biomimetic OPV-based bolaamphiphile (**28**) appending 3′-phosphorylated thymidine moieties at both ends of the rod, and complexed them with complementary 20-meric adenylic acid **29** in the aqueous solution, resulting in A–T base pairs like DNA double

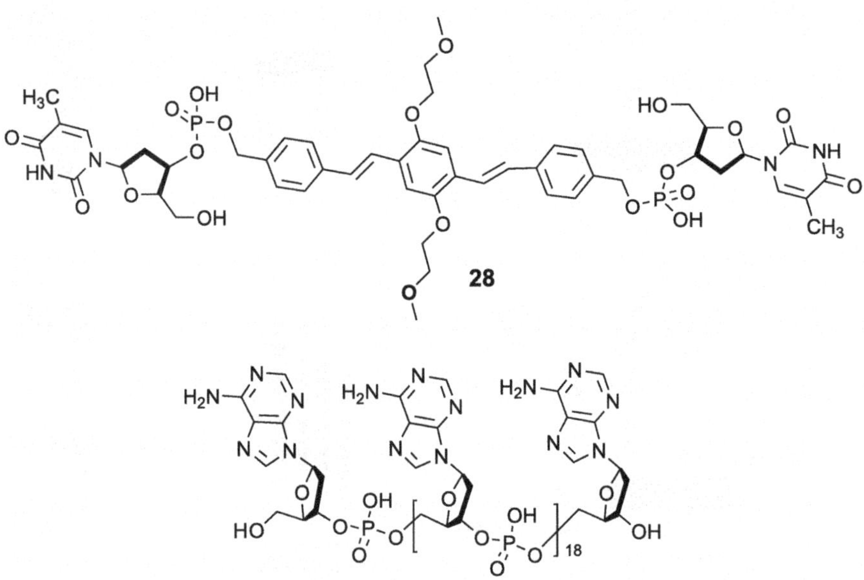

Scheme 15

strands [74]. On the basis of in-depth structural analysis by UV-Vis, CD and AFM techniques, the templating process of **28** with **29** turned out to induce right-handed helical fibers whose widths and lengths could be controlled as a function of T/A ratio in a given complex. The determined diameters of two

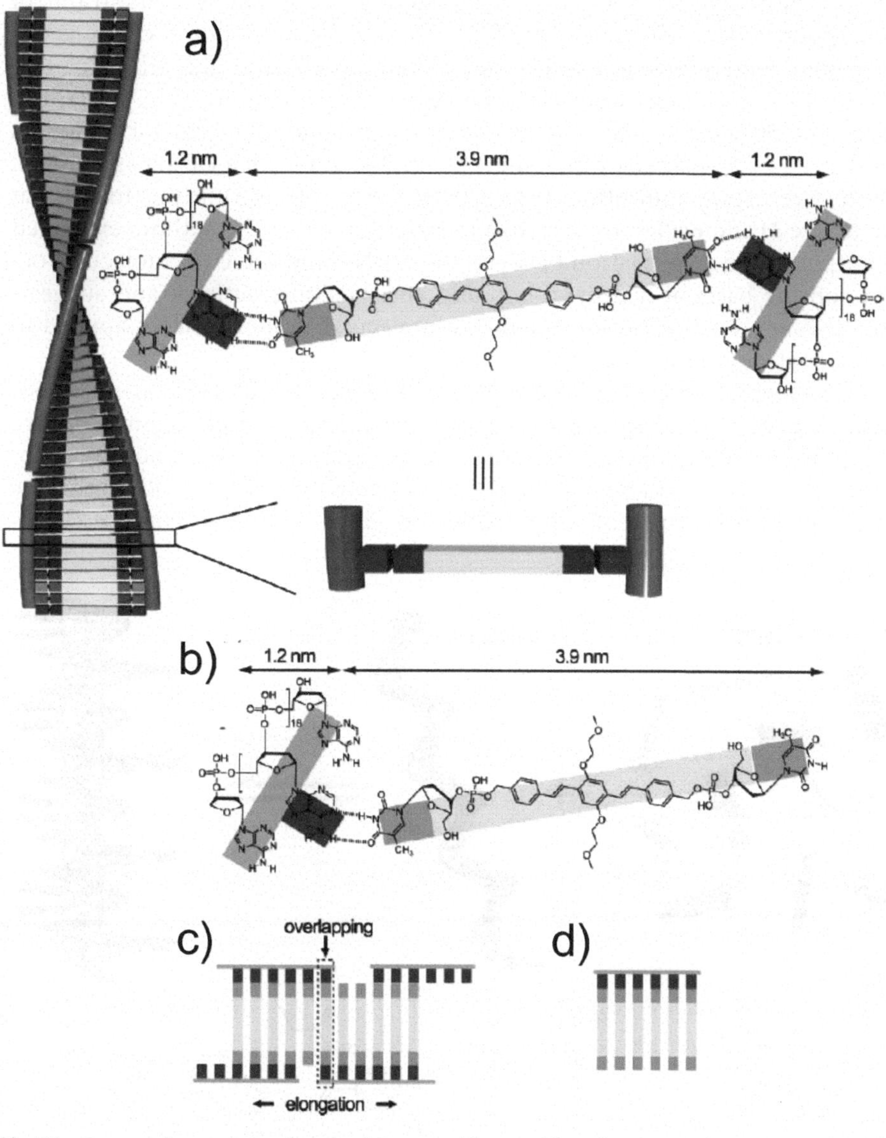

Fig. 25 Proposed structure for the binary self-assembly of **a 28/29** (T : A = 1 : 1) and **b 28/29** (T : A = 2 : 1). Possible elongation mechanism for the helical stacks self-assembled from **c 28/29** (T : A = 1 : 1) and **d 28/29** (T : A = 2 : 1). Reprinted with permission from [74]. © (2006) American Chemical Society

complexes (T/A = 1 : 1 and 2 : 1) from AFM images could be understood by the sum of the length of **28** (3.9 nm) and either one or two **29** (1.2 nm), and the elongated length of the 1 : 1 complex was interpreted by possible partial overlapping of **29** at the end of the stacked **28/29** complex (Fig. 25).

More recently, Schenning et al. employed a single-stranded DNA (ssDNA), i.e., an oligothymine (**30**), as a template for a supramolecular strand of oligo(*p*-phenylene)vinylenes (**31**) [75]. The OPV chromophore contains hydrophilic penta(ethylene oxide)s and a diamino triazine unit which can be bound to the oligothymine via a triple hydrogen-bonding (Fig. 26). Complexes of **30** with $n = 40$ and **31** have a right-handed helical formation as evidenced by a positive Cotton effect above 300 nm from a CD measurement. With electrospray ionization mass spectrometry (ESI-MS), it was found that up to five **31** molecules are attached to **30** with $n = 10$. The authors explained that this specific association between the ssDNA and the OPV molecules occurs on the basis of hydrogen bonding and π–π interactions, and suggests that this approach could be a versatile molecular design concept applied to functional supramolecular electronics.

According to several publications, supramolecular objects bearing carbohydrate groups were shown to act as outstanding multiple ligands toward receptor proteins, and furthermore the shape and size of supramolecular objects are crucial parameters for biological activities [76, 77]. Recently, Lee et al. synthesized rod–coil amphiphiles (**32**–**34**) containing mannose carbo-

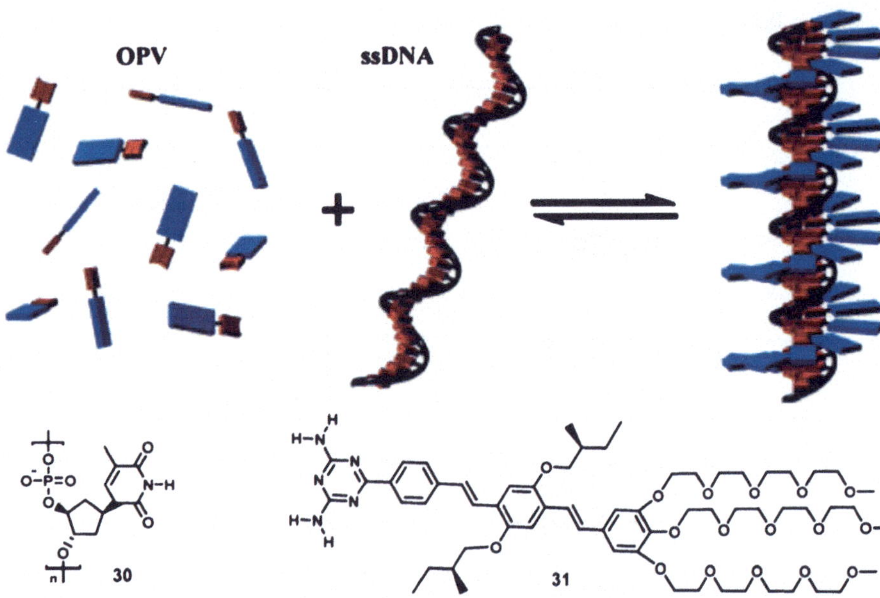

Fig. 26 Schematic illustration of ssDNA templated self-assembly of OPV **31**. Reprinted with permission from [75]. © (2007) American Chemical Society

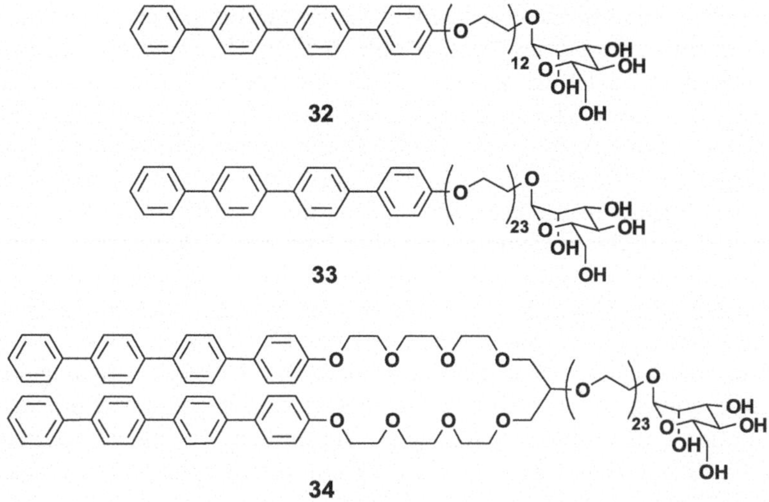

Scheme 16

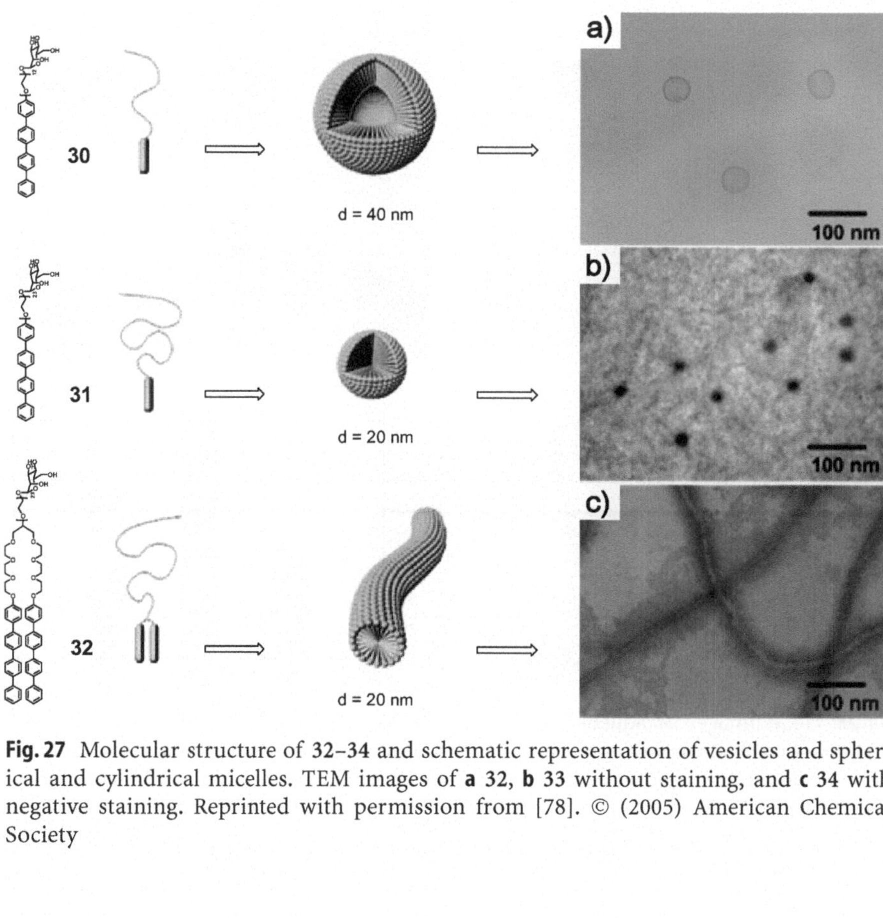

Fig. 27 Molecular structure of 32–34 and schematic representation of vesicles and spherical and cylindrical micelles. TEM images of **a** 32, **b** 33 without staining, and **c** 34 with negative staining. Reprinted with permission from [78]. © (2005) American Chemical Society

hydrates at one end of the coil segment which can bestow enhanced stability and bioactivities on supramolecular objects [78]. The molecular design was implemented by variation of either PEO coil length or rod number. From this molecular approach, the authors fine-tuned the size and shape including vesicles (from **32**), spherical (from **33**) and cylindrical (from **34**) micelles in the aqueous solutions (Fig. 27). The aggregates were observed to act as multivalent ligands in the presence of a natural receptor such as lectin concanavalin A (Con A). From increased object sizes in TEM images, lectin proteins were thought to tightly surround supramolecular objects through multivalent interactions (Fig. 28a–c). Such a specific binding was found exclusively in mannose-based rod–coils; on the other hand, galactose-coated objects did not show the specific object-Con A association. Through a hemagglutination inhibition assay with Con A (to inhibit Con A-promoted erythrocyte agglutination), the authors investigated the influence of object architecture on the binding activity. As a result, they found that spherical micelles with higher curvature from compound **33** showed the highest inhibitory potency (Fig. 28d).

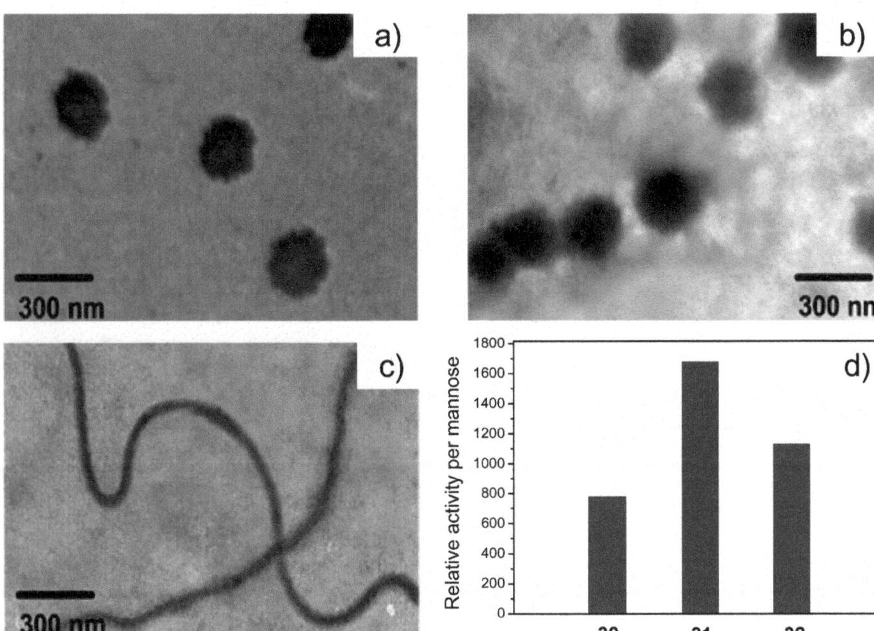

Fig. 28 TEM images of aqueous solutions of **a** 32 and Con A, **b** 33 and Con A, and **c** 34 and Con A at the hemagglutination inhibition concentration. **d** Relative activity of Con A-induced hemagglutination inhibition based on the minimum inhibitory concentration (MIC) of **32–34**. Reprinted with permission from [78]. © (2005) American Chemical Society

More recently, Lee and co-workers also designed a carbohydrate-conjugate molecular wedge (**35**) that can endow aggregates with enhanced stability, external stimuli responsive character as well as biological functions [79]. These molecules were found to self-assemble into nanofibers that could reversibly transform into spheres upon addition of guest aromatic small molecules, i.e. Nile Red (Fig. 29). The structural transformation could be interpreted by the

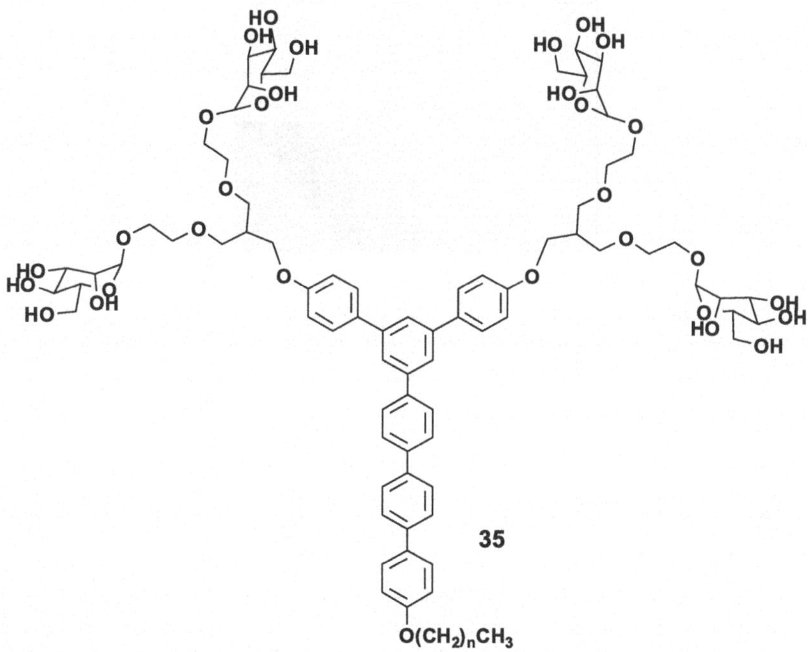

Scheme 17

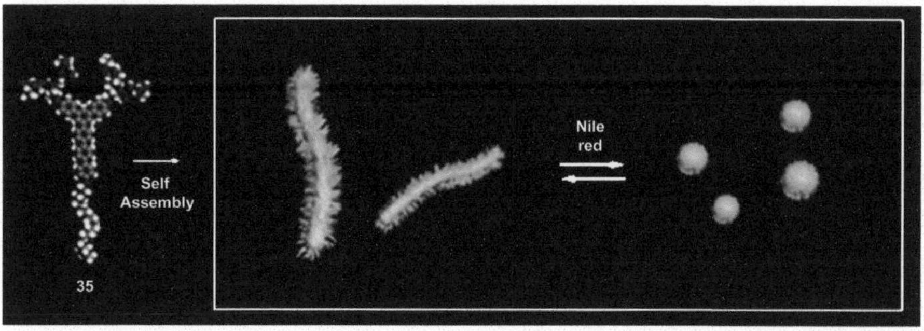

Fig. 29 The schematic representation of the reversible transformation of cylindrical micelles into spherical micelles. Reprinted with permission from [79]. © (2007) American Chemical Society

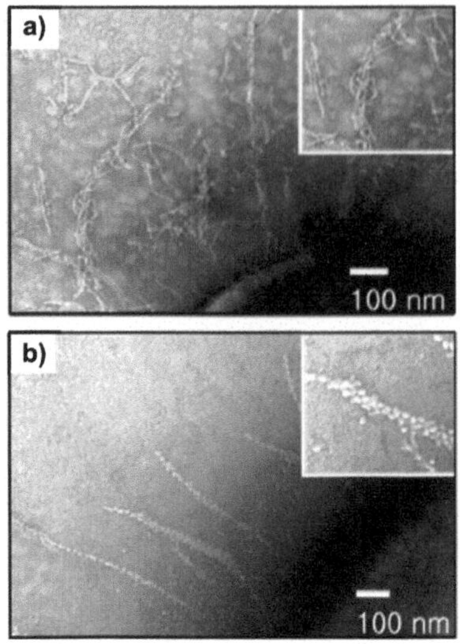

Fig. 30 TEM images with negative staining of a sectioned area of pili of the *E. coli* ORN 178 strain bound with **a** cylindrical (**35**) and **b** spherical micelles (**35**·Nile Red). A portion of an *E. coli* is shown in the *lower right*. Reprinted with permission from [79]. © (2007) American Chemical Society

intercalation of the guest molecules between the aromatic cores, causing the packing of the aromatic wedges to be loose. Moreover, the cylindrical and spherical objects were able to bind to the multiple mannose binding proteins (MBPs) of bacterial pili in *Escherichia coli* (ORN 178) (Fig. 30). Interestingly, the degree of *E. coli* motility turned out to be different depending on the shape of attached supramolecular objects. As a consequence, the authors observed that cylindrical objects inhibit motility of *E. coli* more significantly than spherical objects, and speculated that the cylindrical micelles act as cross-linkers to cause intra-bacterial pili aggregation.

3 Conclusions and Outlook

Molecular rods containing flexible coils have been explored as self-assemblers that could organize spontaneously into diverse nanostructures. Relative to phase-rich bulk-morphologies, self-assembled nanostructures in solutions have not been studied widely. Supramolecular assemblies of rod building blocks in solutions can provide a powerful tool to create functional nanomaterials due to the unique shape and electro-optical functions of rod building blocks. Among the observed solution morphologies, one-dimensional structures such as fibers, tubules and ribbons can be considered as foundations for nanodevices, supramolecular conducting wires, structural scaffolds for gel materials etc.

Rational design of rod amphiphiles into rod–coils, macrocycles, dendron–rod–coils, dumbbells, wedge–coils, and conjugate rods with lateral chains can generate diverse one-dimensional supramolecular architectures responsible for specific material functions. Such a molecular design concept would play a key role in the determination of rod-to-solvent/coil-to-solvent interactions, packing parameters of rods etc. For example, the design of macrocyclic rod amphiphiles in Sect. 2.2, showed a unique formation of tubules with finite tube length, i.e. supramolecular barrels, in the aqueous solution, in contrast to nanotubules with infinite length. In addition, asymmetric dumbbells in Sect. 2.4, displayed an interesting transformation of discrete micelles, supramolecular "rings" into infinite cylinders. Among the structures, the unprecedented toroidal structure molecular rod is an excellent example that shows the importance of the molecular design for tailoring supramolecular assemblies.

As can be observed in natural peptide and DNA strands, in some cases helical features of nanofibers can be obtained by introduction of chiral moieties to rod building blocks. The employment of chiral coils may make it possible to fine tune rod packing structures into helical patterns, leading to supramolecular ribbons or screws. As described in the text, helical strand formation was observed in rod–coils, macrocycles, wedge–coils and conjugate rods with lateral chains. More interestingly, dendron–rod–coils without molecular chirality can self-assemble into a twisted helical structure in certain solvents, which acts as a template for the formation of single- or double-stranded inorganic helices.

Like intelligent robots, the one-dimensional fibers from molecular rods can respond to external stimuli such as small guest molecules. As noted in Sect. 2.4, molecular dumbbells can switch their assembled nanostructures from helical fibers to giant vesicles upon addition/removal of small aromatics. Also, the addition of chiral OPV rods to achiral OPV rods in solution exhibited a similar supramolecular transformation from giant vesicles to helical tubules, as explained by a "sergeant-and-soldiers" coassembly approach (Sect. 2.6). These switchable supramolecular systems, particularly those in water environments, may be utilized as a potent drug delivery system.

The outstanding optical activities of conjugate OPV rods can be upgraded by adopting supramolecular concepts. As described in Sect. 2.6, fibrous aggregates of OPV rod molecules were able to enhance light-harvesting behavior, and further fine-tune emissive colors via donor–acceptor complexes.

It is also interesting that rod assemblies can be applicable to biomimetic or bioconjugate systems. In Sect. 2.7, artificial nucleic acid complexes consisting of amphiphilic rods and linear nucleic acid polymers showed single twisted right-handed helical fibers. In addition, carbohydrate-coated supramolecular objects self-assembled by rod–coils or wedge–coils revealed outstanding ligand activities toward natural receptors such as lectin proteins or bacterial pili of *Escherichia coli*.

Although at present only a few studies related to one-dimensional or the related supramolecular assemblies of rod building blocks have been reported, it is obvious that more intensive research efforts will be made because of the fascinating functionalities of the molecular rod units. Moreover, it is expected that the self-assembling nanostructures would be widely applicable to advanced research areas ranging from material chemistry and nanotechnology all the way up to biological science.

Acknowledgements We gratefully acknowledge financial support from the National Creative Research Initiative Program of the Korea Ministry of Science and Technology, Korea Science and Engineering Foundation (R01-2006-000-11221-0), and a grant from the Fundamental R&D Program for Core Technology of Materials funded by the Ministry of Commerce, Industry and Energy, Republic of Korea.

References

1. Lehn JM (ed) (1995) Supramolecular Chemistry. VCH, Weinheim
2. Berl V, Schmutz M, Krische MJ, Khoury RG, Lehn JM (2002) Chem Eur J 8:1227
3. Ligthart GBWL, Ohkawa H, Sijbesma RP, Meijer EW (2005) J Am Chem Soc 127:810
4. Brunsveld L, Folmer BJB, Meijer EW, Sijbesma RP (2001) Chem Rev 101:4071
5. Sijbesma RP, Beijer FH, Brunsveld L, Folmer BJB, Hirschberg JHKK, Lange RFM, Lowe JKL, Meijer EW (1997) Science 278:1601
6. Lee M, Cho BK, Zin WC (2001) Chem Rev 101:3869
7. Douy A, Gallot B (1987) Polymer 28:147
8. Klok HA, Langenwalter JF, Lecommandoux S (2000) Macromolecules 33:7819
9. Chen JT, Thomas EL, Ober CK, Mao G (1996) Science 273:343
10. Stupp SI, Lebonheur V, Walker K, Li LS, Huggins KE, Keser M, Amstutz A (1997) Science 276:384
11. Tew GN, Pralle MU, Stupp SI (2000) Angew Chem Int Ed 39:517
12. Hoeben FJM, Jonkheijm P, Meijer EW, Schenning APHJ (2005) Chem Rev 105:1491
13. Lee M, Cho BK, Kim H, Lee JY, Zin WC (1998) J Am Chem Soc 120:9168
14. Cho BK, Lee M, Oh NK, Zin WC (2001) J Am Chem Soc 123:9677
15. Cho BK, Chung YW, Lee M (2005) Macromolecules 38:10261
16. He Y, Li Z, Simone P, Lodge TP (2007) J Am Chem Soc 128:2745
17. Zhou Z, Li Z, Ren Y, Hillmyer MA, Lodge TP (2003) J Am Chem Soc 125:10182
18. Chen Z, Cui H, Hales K, Li Z, Qi K, Pochan DJ, Wooley KL (2005) J Am Chem Soc 127:8592
19. Bhargava P, Tu Y, Zheng JX, Xiong H, Quirk RP, Cheng SZD (2007) J Am Chem Soc 129:1113
20. Fuhrhop JH, Helfrich W (1993) Chem Rev 93:1565
21. Behanna HA, Donners JJJM, Gordon AC, Stupp SI (2005) J Am Chem Soc 127:1193
22. Ray S, Das AK, Drewb MGB, Banerjee A (2006) Chem Commun 4230
23. Lohmeijer BGG, Schubert US (2002) Angew Chem Int Ed 41:3825
24. Hamley IW (ed) (1998) The Physics of Block Copolymers. Oxford Science Publications, Oxford

25. Shimizu T, Masuda M, Minamikawa H (2005) Chem Rev 105:1401
26. Gohy JF (2005) Adv Polym Sci 190:65
27. Baddeley C, Yan Z, King G, Woodward PM, Badjic JD (2007) J Org Chem 72:7270
28. Lee HY, Nam SR, Hong JI (2007) J Am Chem Soc 129:1040
29. Estroff LA, Hamilton AD (2004) Chem Rev 104:1201
30. Shenton W, Douglas T, Young M, Stubbs G, Mann S (1999) Adv Mater 11:253
31. Tsukamoto R, Muraoka M, Seki M, Tabata H, Yamashita I (2007) Chem Mater 19:2389
32. Massey JA, Temple K, Cao L, Rharbi Y, Raez J, Winnik MA, Manners I (2000) J Am Chem Soc 122:11577
33. Chécot F, Lecommandoux S, Gnanou Y, Klok HA (2002) Angew Chem Int Ed 41:1339
34. Cornelissen JJLM, Fisher M, Sommerdijk NAJM, Nolte RJM (1998) Science 280:1427
35. Vriezema DM, Hoogboom J, Velonia K, Takazawa K, Christianen PCM, Maan JC, Rowan AE, Nolte RJM (2003) Angew Chem Int Ed 42:772
36. Miyaura N, Suzuki A (1995) Chem Rev 95:2457
37. Deming TJ (2006) Adv Polym Sci 202:1
38. Klok HA (2002) Angew Chem Int Ed 41:1509
39. Kilbinger AFM, Schenning APHJ, Goldoni F, Feast WJ, Meijer EW (2000) J Am Chem Soc 122:1820
40. Schenning APHJ, Kilbinger AFM, Biscarini F, Cavallini M, Cooper HJ, Derrick PJ, Feast WJ, Lazzaroni R, Leclère Ph, McDonell LA, Meijer EW, Meskers SCJ (2002) J Am Chem Soc 124:1269
41. Ryu JH, Lee M (2005) J Am Chem Soc 127:14170
42. Wang H, Wang HH, Urban VS, Littrell KC, Thiyagarajan P, Yu L (2000) J Am Chem Soc 122:6855
43. Wang H, You W, Jiang P, Yu L, Wang HH (2004) Chem Eur J 10:986
44. Sung CH, Kung LR, Hsu CS, Lin TF, Ho RM (2006) Chem Mater 18:352
45. Yang WY, Ahn JH, Yoo YS, Oh NK, Lee M (2005) Nat Mater 4:399
46. Yang WY, Lee E, Lee M (2006) J Am Chem Soc 128:3484
47. Grayson SM, Fréchet JMJ (2001) Chem Rev 101:3819
48. Zubarev ER, Pralle MU, Sone ED, Stupp SI (2001) J Am Chem Soc 123:4105
49. De Gans BJ, Wiegand S, Zubarev ER, Stupp SI (2002) J Phys Chem B 106:9730
50. Zubarev ER, Sone ED, Stupp SI (2006) Chem Eur J 12:7313
51. Zubarev ER, Pralle MU, Sone ED, Stupp SI (2002) Adv Mater 14:198
52. Sone ED, Zubarev ER, Stupp SI (2002) Angew Chem Int Ed 41:1706
53. Sone ED, Zubarev ER, Stupp SI (2005) Small 1:694
54. Bae J, Choi JH, Yoo YS, Oh NK, Kim BS, Lee M (2005) J Am Chem Soc 127:9668
55. Ryu JH, Kim HJ, Huang Z, Lee E, Lee M (2006) Angew Chem Int Ed 45:5304
56. Kim JK, Lee E, Huang Z, Lee M (2006) J Am Chem Soc 128:14022
57. Zhang X, Chen Z, Würthner F (2007) J Am Chem Soc 129:4886
58. Israelachvili JN (ed) (1985) Intermolecular and Surface Forces. Academic Press, London
59. Kim JK, Lee E, Lee M (2006) Angew Chem Int Ed 45:7195
60. Jonkheijm P, Miura A, Zdanowska M, Hoeben FJM, De Feyter S, Schenning APHJ, De Schryver FC, Meijer EW (2004) Angew Chem Int Ed 43:74
61. Schwab PFH, Smith JR, Michl J (2005) Chem Rev 105:1197
62. Ajayaghosh A, Praveen VK (2007) Acc Chem Res 40:644
63. Ajayaghosh A, George SJ (2001) J Am Chem Soc 123:5148
64. Ajayaghosh A, George SJ, Praveen VK (2003) Angew Chem Int Ed 42:332
65. Ajayaghosh A, Vijayakumar C, Praveen VK, Babu SS, Varghese R (2006) J Am Chem Soc 128:7174

66. Ajayaghosh A, Praveen VK, Vijayakumar C, George SJ (2007) Angew Chem Int Ed 46:6260
67. George SJ, Ajayaghosh A, Jonkheijm P, Schenning APHJ, Meijer EW (2004) Angew Chem Int Ed 43:3422
68. Ajayaghosh A, Varghese R, Mahesh S, Praveen VK (2006) Angew Chem Int Ed 45:7729
69. Ajayaghosh A, Vijayakumar C, Varghese R, George SJ (2006) Angew Chem Int Ed 45:456
70. Miller RA, Presley AD, Francis MB (2007) J Am Chem Soc 129:3104
71. Bronco S, Cappelli C, Monti S (2004) J Phys Chem B 108:10101
72. Niu Z, Bruckman MA, Li S, Lee LA, Lee B, Pingali SV, Thiyagarajan P, Wang Q (2007) Langmuir 23:6719
73. Sleytr UB, Messner P, Pum D, Sára M (1999) Angew Chem Int Ed 38:1034
74. Iwaura R, Hoeben FJM, Masuda M, Schenning APHJ, Meijer EW, Shimizu T (2006) J Am Chem Soc 128:13298
75. Janssen PGA, Vandenbergh J, van Dongen JLJ, Meijer EW, Schenning APHJ (2007) J Am Chem Soc 129:6078
76. Thoma G, Katopodis AG, Voelcker N, Duthaler RO, Streiff MB (2002) Angew Chem Int Ed 41:3195
77. Aoyama Y, Kanamori T, Nakai T, Sasaki T, Horiuchi S, Sando S, Niidome T (2003) J Am Chem Soc 125:3455
78. Kim BS, Hong DJ, Bae J, Lee M (2005) J Am Chem Soc 127:16333
79. Ryu JH, Lee E, Lim YB, Lee M (2007) J Am Chem Soc 129:4808

Adv Polym Sci (2008) 219: 107–171
DOI 10.1007/12_2008_146

Published online: 20 May 2008

Functional Self-Assembled Nanofibers by Electrospinning

A. Greiner · J. H. Wendorff (✉)

Department of Chemistry and Center of Material Science, Philipps-University, 35032 Marburg, Germany
wendorff@staff.uni-marburg.de

1	**Introduction**	108
2	**Nature of the Electrospinning Process**	114
2.1	Experimental Setups	114
2.2	Experimental Observations on Fiber Formation	115
2.2.1	The Straight Path of the Jet	117
2.2.2	The Looping Part of the Jet	117
2.3	Theoretical Analysis	120
2.3.1	Droplet Deformation and Jet Initiation	120
3	**Nanofiber Properties**	128
3.1	Nanofiber Diameters	128
3.2	Shape of the Fibers	131
3.3	Nanofiber Topologies, Porous Fibers	134
3.4	Internal Morphology	137
3.4.1	Amorphous Polymers	138
3.4.2	Partial Crystalline Nanofibers	139
3.5	Mechanical Properties of Single Nanofibers	141
4	**Nonwovens Composed of Electrospun Nanofibers**	142
4.1	Fiber Arrangement in Nonwovens	143
4.2	Heterogeneous Nonwovens	144
4.3	Porosity and Pore Structures	146
4.4	Mechanical Properties of Nonwovens	148
5	**Recent Electrospinning Developments**	149
5.1	Electrospinning with Strongly Reduced Electrode Distances	149
5.2	Co-Electrospinning	151
5.3	Core Shell Fibers and Hollow Fibers via Templates (TUFT Approach)	152
6	**Brief Review of Materials Which have been Electrospun to Fibers**	154
6.1	Spinning of Technical Polymers	154
6.1.1	Spinning from Polymer Melts	154
6.1.2	Electrospinning from Organic Solvents	156
6.1.3	Electrospinning of Water Soluble Polymers	157
6.2	Spinning of Biopolymers	158
6.3	Nanofibers from Polymer Hybrids, Metals, Metal Oxides	160
6.3.1	Polymer Hybrids	160
6.3.2	Metal and Metal Oxide Nanofibers	162

7	**Applications for Electrospun Nanofibers**	164
7.1	Technical Applications	164
7.1.1	Template Applications	164
7.1.2	Textile Applications	164
7.1.3	Filter Applications	165
7.1.4	Catalysis	165
7.1.5	Nanofiber Reinforcement	165
7.2	Medicinal Applications	166
7.2.1	Tissue Engineering	166
7.2.2	Wound Healing with Nanofibers	166
7.2.3	Transport and Release of Drugs/Drug Delivery	167
	References	168

Abstract Electrospinning constitutes a unique technique for the production of nanofibers with diameters down to the range of a few nanometers. In strong contrast to conventional fiber producing techniques, it relies on self-assembly processes driven by the Coulomb interactions between charged elements of the fluids to be spun to nanofibers. The transition from a macroscopic fluid object such as a droplet emerging from a die to solid nanofibers is controlled by a set of complex physical instability processes. They give rise to extremely high extensional deformations and strain rates during fiber formation causing among others a high orientational order in the nanofibers as well as enhanced mechanical properties. Electrospinning is predominantly applied to polymer based materials including natural and synthetic polymers, but, more recently, its use has been extended towards the production of metal, ceramic and glass nanofibers exploiting precursor routes. The nanofibers can be functionalized during electrospinning by introducing pores, fractal surfaces, by incorporating functional elements such as catalysts, quantum dots, drugs, enzymes or even bacteria. The production of individual fibers, random nonwovens, or orientationally highly ordered nonwovens is achieved by an appropriate selection of electrode configurations. Broad areas of application exist in Material and Life Sciences for such nanofibers, including not only optoelectronics, sensorics, catalysis, textiles, high efficiency filters, fiber reinforcement but also tissue engineering, drug delivery, and wound healing. The basic electrospinning process has more recently been extended towards compound co-electrospinning and precision deposition electrospinning to further broaden accessible fiber architectures and potential areas of application.

Keywords Co-electrospinning · Electrospinning · Fiber architectures · Functions and applications · Nanofibers · Nonwovens · Precision electrospinning

1 Introduction

Spider webs are impressive for their complex architecture that gives rise to specific functions and in particular for their high performance ultrafine functionalized fibers from which the webs are constructed [1]. Several different types of silk are being used in web construction, including a "sticky" capture silk or "fluffy" capture silk, depending on the type of spider. Silk obtained from cocoons made by the larvae of the silkworm Bombyx mori has a shimmering appearance which originates from the triangular prism-like structure of the fibers. This allows silk cloth to refract incoming light at different angles. Silk fibers possess highly impressive mechanical properties, in particular a high ductility, related to intrinsic structural features again being ultrafine in nature. Nature proves by these examples and many more not discussed here to be extremely efficient in creating functional materials in the shape of fine fibers.

This also holds for natural fibers such as cotton, wool, hairs, etc. [2]. These fibers are not as small in fiber diameter as the ones discussed above yet they are constructed in a highly complex hierarchical way which provides them not only with unique mechanical properties but also with another set functions which make them of interest for various types of applications. The diameter of such natural fibers may well be in the 10–20 μm range and above; human hair, for example, typically has a diameter around 50 μm. Strong correlations exist between a variety of functions displayed by natural fibers and their molecular and supermolecular structure devised by nature. Natural fibers composed of silk, wool, or cotton are used predominantly for textile applications providing functions such as thermal insulation, wind resistance, exchange of water vapor, etc., but they also contribute in unique ways in fashion design which frequently relies on the optical effects of silk produced by the prism-shaped fibers.

Man-made fibers, composed of materials such as polyamides or polyethylene terephthalate and produced via synthetic routes, have to a significant extent replaced natural fibers in textiles. They are advantageous, because they are cheaper to produce and easier to process, dye, or introduce high strength and stiffness in a controlled way [3]. Yet, to a certain extent, they miss quite a number of functions that are beneficial for textile applications and which are displayed by natural fibers. Wearing a shirt made purely from polyamide on a hot humid day makes the difference between textiles for example, from cotton and from man-made fibers very obvious. One major reason besides a chemical composition that is different from the one of the natural fibers discussed above is that their internal molecular and supermolecular structures tend to be rather simple. Solid fibers with constant composition and constant structural features along the cross section are characteristic of man-made fibers. Hollow fibers and also fibers with cross-sectional shapes that differ from the circular one have been produced, yet nevertheless such architectures are far from the complex ones displayed by nature [4]. Furthermore, specific microfibers have been produced along various ways, to enhance among other things, textile properties. A further strong reduction of the diameter of fibers used in textiles would at least greatly enhance thermal insulation, wind resistance, moisture absorption, exchange of vapor etc. It even seems possible that complex fiber morphologies may become accessible through control by confinement effects.

In fact, the application of man-made fibers is not only restricted to textiles. Fibers play a major role in the reinforcement of thermoplastic and thermoset polymer matrices for high-end applications [3, 5]. Reinforcement of high end elements of ships, trains, and airplanes are well known, but fiber reinforced materials can also be found in day to day appliances. As far as fiber reinforcement is concerned, the important parameters are the axial ratio, which should be well above 100 to 1000 (thus the use of fibers), and the enhanced stiffness and strength of the fiber combined with a good mechanical coupling to the matrix. Fiber reinforcement is in the majority of cases not done with natural fibers [2, 3], although such approaches are being considered more and more for ecological reasons, but is conventionally done by using very specific synthetic fibers (such as carbon fibers produced among others via precursor polymer polyacrylonitrile (PAN) fibers or Kevlar fibers, produced from lyotropic polymer solutions) [3]. Fiber reinforcement could stongly benefit from fibers much smaller in diameter, even at the same magnitudes of stiffness and strength, since the length could be reduced at constant axial ratio compared to thicker fibers reducing thus rupture during polymer processing. In addition, fibers small in diameter compared to the wavelength of light would not cause turbidity in otherwise transparent matrices. It follows that the mechanical coupling between matrix and fibers will be enhanced, and, thus, the ductility, due to the much larger internal surface areas.

A further area for the application of fibers, again predominantly of man-made fibers, concerns filters for either gas or fluid filtrations including coalescer filters [6]. The chemical, thermal, and mechanical stability of the fibers together with the costs to produce the fibers are important features, but the absolute magnitude of the diameter of the fibers is of particular importance. The diameter determines the size of the pores provided by the filters and thus the size of the impurities to be filtered out. The reduction to fiber diameters within the nanometer range will affect the flow pattern around the fibers significantly and should strongly enhance the filter efficiency with respect to smaller scale impurities in the air, in gasoline, etc. The specific surface area acting as adsorption site also increases strongly as the diameter is reduced.

The discussion about fiber applications in the areas of textiles, fiber reinforcement, and filters has made it apparent that these areas would benefit to a great extent from a further strong reduction of the fiber diameters by several orders of magnitude well into the nanometer range. The low value of the diameter and small nonwoven pore sizes, as well as the huge surface area which goes along with small fiber diameters, are key factors in such applications. Yet, it is obvious that the extremely small diameter is just one side of the coin. Further features are the onset of confinement effects for structural features and properties and the increasing truly 1-dimensional nature of the fibers as the diameter decreases. The potential for rapid diffusional processes into and out of the fiber characteristic of nanoscalar dimensions, the close resemblance in architecture of electrospun fibers, and the fibrillar extracellular matrix in living systems are further specific features favorable to specific applications. It may, of course, be necessary for such fibers to carry functional units such as chromophores, catalysts, sensor molecules, quantum dots, drugs, or bacteria depending on the application in mind, and they may have to be composed of organic, inorganic materials, or corresponding hybrids.

Conventional processing techniques will not be able to yield such extremely fine functionalized fibers. This inability also holds for melt blowing and similar special techniques [7, 8]. The technique of choice is electrospinning. Prior to the year 2000 electrospinning was the domain of a few specialists; the average number of papers published per year on this topic was well below 20. This situation has changed dramatically in the last few years. In 2007, more than 500 papers have been published on electrospinning. It is estimated that more than 200 research groups in academia and industry work currently on this topic, and the number of conferences and conference sessions devoted to electrospinning is continuously increasing. Electrospinning has become a widely appreciated nanostructuring technique in academia and industry and, in fact, has a lot to offer [9–13].

Basically it allows production of nanofibers with diameters down to a few nanometers from a broad range of polymers. Yet, due to the unique self-assembly processes happening in electrospinning, it is a highly versatile technique in terms of the materials that can be spun to nanofibers, the control of their morphology, their surface topology, as well as the properties of the fibers and nonwovens composed of them. Figure 1 illustrates the broad range of fiber architectures available from electrospinning including thin smooth fibers, porous fibers, and fibers with fractal surface structures, with spindle-type disturbances, ribbonlike fibers or odd-shaped fibers, such as "barbed" nanowires.

A multitude of functions can be incorporated into these fibers, and an extremely broad range of potential applications exists in which electrospun fibers can make major contributions. These include not only textile, filter and mechanical reinforcement applications but also extend to tissue engineering, drug delivery, wound healing, sensorics, optoelectronics, catalysis, and many more applications. The progress achieved in electrospinning in a time span covering less than one decade, coupled with the strong impact it has made and continues to make on Material and Life Science are unique features. A set of review articles have recently been published that provide an insight into the vast opportunities afforded by electrospinning [9–13].

It is of particular importance for the discussions that follow to point out that fiber formation processes in electrospinning differ fundamentally from the ones in conventional technical approaches, such as extrusion and subsequent elongation, melt blowing, or even techniques exploiting converging flow, which all involve mechanical forces and geometric boundary conditions [14, 15]. In melt or solution extrusion the shape and diameter of the die, as well as mechanical forces inducing specific draw ratios and drawing speeds, to a major extent

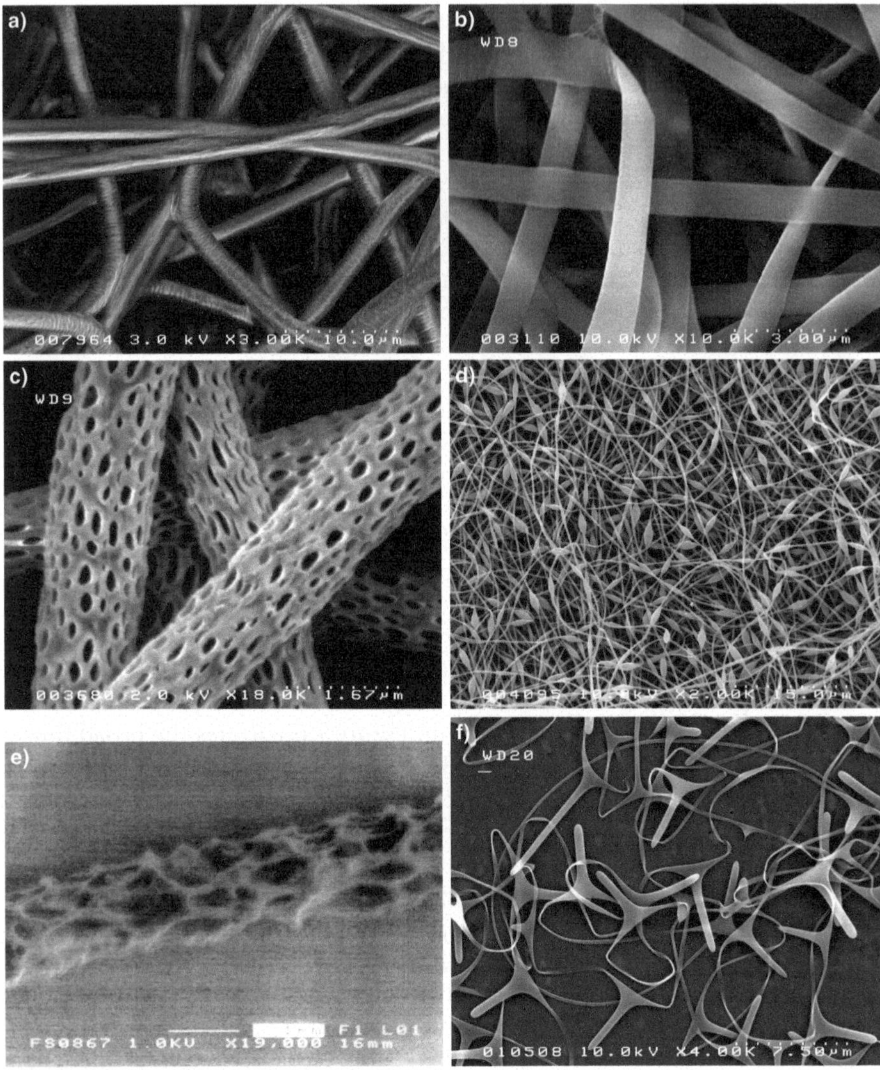

Fig. 1 Nanofiber architectures available from electrospinning: **a** Thin smooth fibers with circular cross section (PA 6), **b** Ribbonlike fibers (PA 6) **c** Porous fibers (PLA) and **d** Fibers with spindle-type disturbances (PAN), **e** Fibers with fractile shape (PLA), **f** "Barbed" nanofibers (PVA)

determine dimensional and structural properties of the resulting fibers. Fiber formation is thus controlled by mechanical deformation processes to which the original solution or melt are subjected. In an extension of the extrusion technique, multicomponent fibers consisting of segments of different polymers can be fabricated by extrusion. The subsequent preparation of fine fibers is induced by splitting up the fiber by treatment with, for example, water jets [8]. The melt blowing process is a technically advanced process leading to fibers with diameters below 500 nm. In the melt-blown technology polymer melts are pumped through an array of nozzles. In the process, the formation of fibers from the melt is obtained via cooling in a strong countercurrent of air again imposing specific mechanical forces.

Another interesting approach is the exploitation of converging flow to first produce droplets, as in electrospraying, but in specific cases, including nanofibers [14, 15]. The concept in one particular setup is to start from a fluid layer arrangement composed of two immiscible fluids, one of which serves as processing fluid and the other constituting the material to be processed to fibers. These fluid layers are sucked by mechanical forces through a die which imposes a converging flow. For slow sucking speeds only the upper sacrificial fluid layer is subjected to the converging flow, yet at higher speeds the lower fluid layer is also sucked in, yielding a compound jet with the sacrificial material forming the outer shell. The fluid core fiber is subsequently either subjected to a breakup, yielding droplets, or solidifies, yielding fibers with diameters well below the micrometer range. Using three layer arrangements, core shell fibers or hollow fibers can be produced in this way. Again fiber formation is controlled in this approach by mechanical forces as in conventional extrusion, although the strong correlation existing between die diameter and final fiber diameter is relaxed to a certain extent.

Fiber formation in electrospinning differs strongly from the formations occurring in the approaches discussed so far, since self-assembly processes dominate in electrospinning (which will become apparent in detail as the theory of electrospinning is explained below). Features that are basically governed by self-assembly processes induced by specific electrostatic interactions of elements of the original source droplet or similar geometries from which fiber formation starts include: the evolution of the final diameter of the nanofibers resulting from electrospinning, the intrinsic orientational order, the morphology, the cross-sectional shape, gradients along the cross section, specific phase morphologies, as well as the distribution of solid particles dispersed within the fiber, undulations of the fiber diameter, and droplets arranged along the fibers in a regular fashion.

In the case of supramolecular structure formation self-assembly is known to be controlled by specific, generally attractive, forces such as hydrogen bonding, charge transfer interactions, etc. [16, 17]. Self-assembly in electrospinning, on the other hand, is controlled by Coulomb interactions between charged elements of the fluid body. Self-assembly follows the general Earnshaw theorem of electrostatics according to which it is impossible to prepare stable fluid structures such as stable fluid jets in which all elements interact only by Coulomb forces [18, 19]. Charges located within the fluid jet, in the case considered here, move the polymer elements to which they are attached along complex pathways in such a way that the Coulomb interaction energy is minimized. Droplet deformation, jet initiation, and, in particular, the bending instabilities that to a major extent control fibers' properties are apparently predominantly controlled by this kind of self-assembly principle. Simple as well as highly complex fiber architectures, the deposition of flat fibers, of fibers with vertical protrusions, and splayed fibers can be considered as a manifestation of the variety of self-assembly processes happening in electrospinning (see examples in Fig. 1). Keeping this general theorem in mind it should be possible to come up with completely novel fiber architectures – Fig. 1f may serve as an example – and thus novel functions in the future. In view of this very specific character of fiber formation via electrospinning, it seems reasonable to first consider the electrospinning process in detail, both from an experimental and theoretical point of view, before embarking on the Material Science aspects and application sides of electrospinning.

2 Nature of the Electrospinning Process

2.1 Experimental Setups

At first glance, electrospinning gives the impression to be a very simple and therefore easily controllable technique to produce fibers with dimensions down into the nanometer range. In electrospinning on a simple laboratory scale, a polymer solution or melt is pumped through a thin capillary acting as a die with inner diameters in the range of some 100 μm (Fig. 2). The die may simultaneously serve as an electrode to which high electric fields in the range of typically 100 to 500 kV/m are applied. The distance to the counter electrodes typically amounts to 10 up to 25 cm in laboratory systems and currents flowing during electrospinning are in the range of some 100 nanoamperes up to several microamperes. The substrate on which the electrospun fibers are collected is either the counter electrode material itself or specific substrates selected in view of the target application. It may be brought into electric contact with the counter electrode but it can also be on a different potential. Frequently a top-down arrangement of the die/electrode and the counter electrode is used (Fig. 2), but in principle electrospinning can be carried out with the jet flowing from bottom to top, top to bottom, and horizontally.

To discuss other types of electrospinning devices one has to take into account the fact that electrically initiated jets will generally not start easily on flat fluid surfaces but rather at protrusions [20, 21]. Such protrusions can have the shapes of pending droplets located at the tip of syringe-like dies, an approach used typically in laboratory style electrospinning device, or of sessile droplets positioned on flat solid surfaces [21]. In more technically oriented spinning devices such protrusions may also be formed via metal spikes arranged along

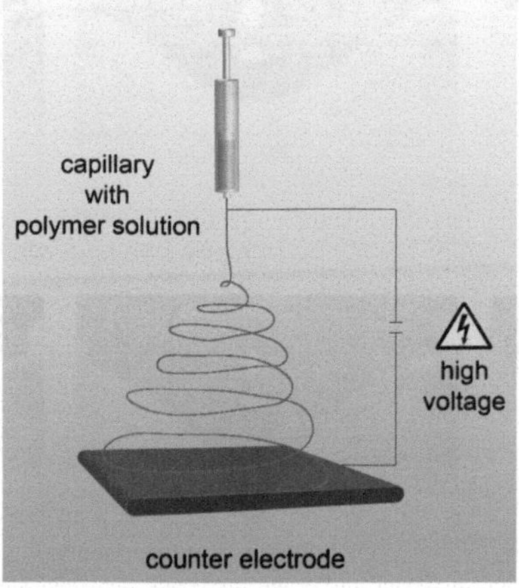

Fig. 2 Electrospinning device, laboratory scale, schematic representation

metal wires or along metal cylinders immersed into or carrying the spinning fluid. The spikes thus assume two roles: they act as solution feeding elements as well as initiation elements. Another unique approach towards the creation of protrusion on fluid surfaces consists in inducing statistical surface roughness modulations, for instance via superparamagnetic particles immersed in the spinning solution via their interactions with magnetic fields [22]. Magnetic fields tend to induce spike structures for such systems. Many more approaches along this line seem feasible and possibly necessary if one intends to scale up the production rate considerably. Multiple die arrangements have been used to this end with limited success. One reason is that the equally charged jets emanating from these dies tend to reject each other and that such an arrangement can only be optimized to deposit fibers on a substrate in a completely homogeneous way with great difficulties.

2.2
Experimental Observations on Fiber Formation

Experimental observations employing, for example, high-speed video analysis reveal for electrospinning devices of the kind introduced above (i.e., for laboratory scales) a sequence of complex fiber forming processes resulting finally in the deposition of extremely fine fibers, of nanofibers, on the counter electrode or substrates such as glass, silica, filter paper or textiles located on top of the counter electrode. The first step towards fiber formation consists in the deformation of fluid drops emerging from the die by the interaction of the applied field with the charged fluid [23] (Fig. 3). With increasing electric field, the shape of the droplet becomes increasingly prolate and approaches a con-

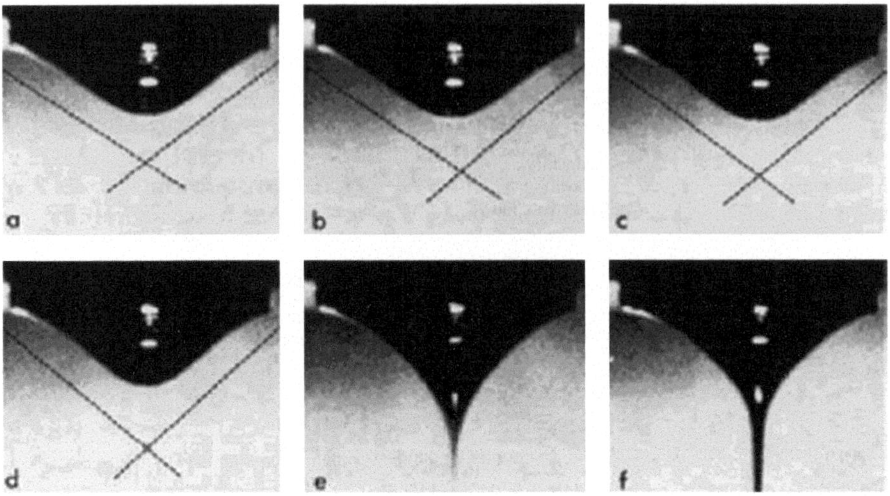

Fig. 3 Deformation of a pending droplet by electric fields [23] (see discussion below)

ical shape with the half angle of the cone assuming values of the order of 30° (Fig. 3a–d). The cone is furthermore characterized by a tip with a very low radius of curvature, well below 1 μm, and thus not readily resolvable by optical means [19–21].

A fluid jet emanates from this tip as a critical electric field is surpassed (Fig. 3e and f). This jet moves towards the counter electrode in a linear fashion for a short distance amounting typically to several centimeters. At the end of the straight path unstable bending motions occur with growing amplitudes and the jet begins to follow a spiralling and looping path in space (Fig. 4).

Fig. 4 Path of a fluid jet in electrospinning, schematic representation as adapted from [19]

2.2.1 The Straight Path of the Jet

In the following, experimental observations on dynamics related to this straight path of the jet collected from tracer particle tracking techniques based on high-speed photography will be discussed [13, 24]. The fluid jet is found to experience a very strong acceleration as it leaves the die going up to 600 m/s^2, which is close to two orders of magnitude larger than the acceleration coming from gravitational forces. Gravitational forces thus play no significant role in electrospinning. This is the reason why top-down, bottom-up, and other types of die/counter electrodes work similarly. The velocity of the jet amounts typically to 3–5 m/s at the end of the straight part of the jet. Further characteristic values are a strain rate which goes up to values of the order of 1000 s^{-1} and an elongational deformation approaching values of up to 1000. An important fact is that an increase of the voltage causes on one side the jet diameter to increase, whereas with the acceleration, the jet velocity as well as the strain rates decrease with significantly increased voltage. The interpretation is that a lower voltage gives rise to a lower feeding rate of the fluid, which in turn causes the jet to be thinner thus allowing larger surface charge densities. The surface charge density is a controlling parameter in fiber formation via electrospinning.

The strain rates characteristic for the straight part of the jet are sufficiently large enough to induce chain extension. Arguments on requirements to be met by the strain rate relative to the hydrodynamic relaxation in order to induce chain orientation have been put forward by de Gennes [25]. These lead to the conclusion that the product of the viscoelastic relaxation time and the strain rate should be larger than 0.5 for the induction of chain extensions. In fact, this product goes up to 50 in the straight path of the jet and again it goes down significantly – to 30 and below – as the electric field is increased, e.g., from approximately 50 to approximately 70 V/mm. It is thus not surprising that, in fact, birefringence has been observed for this part of the jet, which is still, of course, fluid. A further observation is that the birefringence tends to be stronger at the surface of the jets. Some reasons are that the surface charges are located just there, that polymer chains located at the surface possess lower degrees of freedom which tends to make them more susceptible to deformation, and that the surface layers tend to have a higher concentration of polymer chains due to the evaporation of the solvent.

2.2.2 The Looping Part of the Jet

At some distance away from the die the jet is no longer able to follow a straight path in the direction of the counter electrode. It bends, turns sideways, and begins to perform spiralling, looping motions. In each loop the jet becomes thinner and elongated as the loop diameter increases. The envelope of these loops, which is apparent in electrospinning to the naked eye, resembles a cone with its opening oriented towards the counter electrode. This type of instability – bending (also known as whipping instability) – repeats itself in a self-similar fashion on a smaller and smaller scale as the jet diameter is reduced, contributing to a further reduction of the jet diameter. Finally, the jet is so thin or the fiber becomes so stiff that the bending instability can no longer govern the fiber formation process. The deposition of solidified nanofibers onto the counter electrode or substrates located on top of the counter electrode is the final step in electrospinning. The radius of the envelope cone which typically assumes a value of the order of 10 up to 15 cm over a characteristic distance between the die and the counter electrode of approximately 10 to 15 cm also controls the radius of the planar nonwoven, which is deposited in the plane of the counter electrode of similar magnitude.

Nonwovens are defined in this context as textiles which are neither woven nor knit, for example felt. Nonwoven fabric is typically manufactured by putting small fibers together in the form of a sheet or web. Note that in electrospinning the deposition texture depends on the electrode configurations. In the case of planar counter electrodes, a planar texture results, i.e., the fibers are randomly oriented within the plane of the substrate as shown in Fig. 5 for the case of fibers from polyvinyl acetate (PVA). A highly porous nonwoven results with the sizes of the pores, which is on the average much larger than the diameter of the fibers.

Both the elongation and thinning of the fiber within the linear pathway and within the looped pathway are accompanied by solvent evaporation, contributing also to jet diameter thinning and to the final diameter of the solid nanofibers deposited on the counter electrodes. Again using high-speed video analysis and laser Doppler velocimetry, the magnitudes of the total elongation of the jet during electrospinning, the deformation rate, and the speed with

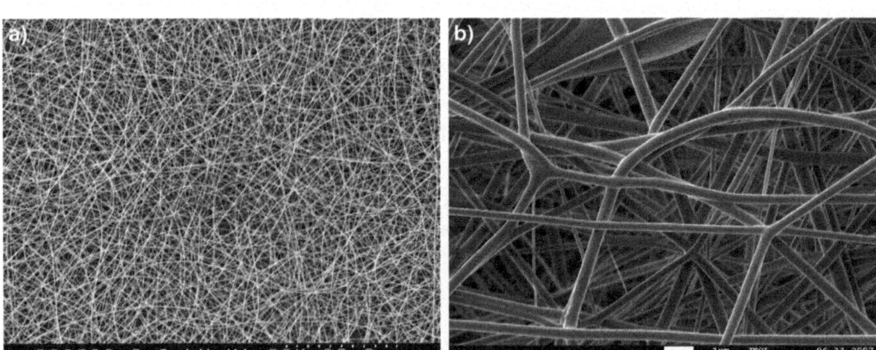

Fig. 5 Nanofiber nonwoven obtained by electrospinning shown at two different magnifications

which the fibers are deposited on the counter electrode can be estimated [26]. The total time which a fluid element experiences within the fiber formation process from leaving the die until becoming deposited on the counter electrode as element of the solid nanofiber is estimated to be approximately 0.2 s. The overall draw ratio is estimated to be of the order of 10^5 and the overall strain rate to be of the order of $10^5\ s^{-1}$ [13, 19]. These are extremely high values reflecting the reduction of the jet diameter from around 100 μm down to 100 nm and below in a very short time (well below a second), when taking into account the evaporation of the solvent and the corresponding reduction in diameter. It is obvious that chain molecules tend to become highly oriented in the jet during electrospinning and that this should show up in the orientational order within the final solid fibers as well as in the crystal morphology. Furthermore, such a strong mechanical deformation should also be reflected in the stiffness and strain of the fibers (these topics will all be discussed later in some detail).

In general solid fibers are deposited on the counter electrodes despite the short fiber formation times well below a second and even if solvents with high boiling points, such as water, are used for the spinning solutions. Yet, depending on the relative humidity, high solvent concentrations, and the use of polymer materials with glass transition temperatures close to or below room temperature, soft fibers are deposited giving rise to partial coalescence effects (Fig. 6a). Coalescence processes set in whenever two soft fibers within the fiber network resulting from electrospinning come in close contact or whenever two such nanofibers cross each other, respectively [89]. Coalescence of fluid droplets in close contact and also of elliptically deformed droplets have been studied widely experimentally as well as on a theoretical basis, yet it seems that such calculations were not extended to fiber coalescence in nonwovens.

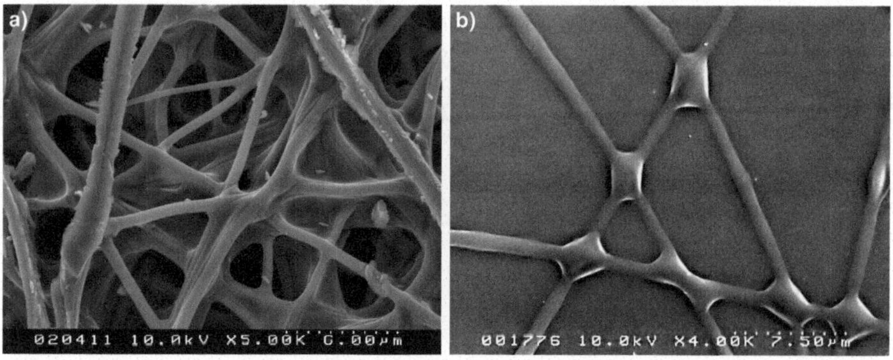

Fig. 6 Coalescence of soft nanofibers. **a** Directly during deposition of soft fibers in electrospinning of cellulose acetate, **b** Result of annealing of polylactide nanofibers at enhanced temperatures [89]

The observation is that coalescence sets in at the contact points of the nanofibers yielding distinct geometries for the junction points, which depend on the angle with which the fibers contact each other and also on the diameters of the two fibers in question, i.e., whether they have a similar diameter or different diameters. The common feature of these geometries is that the disrupted directional variations characteristic of the crossing of two fibers that have still retained their circular cross section is replaced by geometries characterized by more continuous variations of the curvature. The coalescence figures that result during electrospinning are frequently kinetically controlled, as governed by the simultaneous processes of jet deposition and solidification due to solvent evaporation. To obtain coalescence figures corresponding more closely to the equilibrium one may anneal electrospun nanofibers crossing each other at elevated temperatures above the glass or melting temperatures for some minutes [89]. The geometries which result are displayed in Fig. 6b. They quite obviously correspond to low surface free energy configurations spontaneously produced by the tendency of the system to approach the lowest state of surface free energy. Such coalescence processes lead to a mechanical cross-linking of the nonwovens. Such features may be of interest for a set of applications including filter, textile, or tissue engineering applications. Cross-linking may enhance the stiffness and strength of the nonwovens, and cross-linked nonwovens tend to keep their integrity even in the presence of mechanical forces.

It is helpful at the end of this section to compare specific textile fiber properties of conventionally produced fibers with diameters in the micrometer range with fibers in the nanometer range that are produced by electrospinning. From one gram of polyethylene fibers, a total length of 13 km can be produced if the fiber diameter is 10 μm, but a length of 130 000 km if the diameter is 100 nm. In the first case the specific surface, which is the surface given in m^2 per gram fibers, amounts to about 0.4 m^2/g while in the second case it amounts to 40 m^2/g. In fiber technology the unit denier is often used of as a measure of fiber fineness. It determines the mass of a fiber with a length of 9000 m. For a 10 μm fiber the fineness amounts to 1 denier, for a 100 nm fiber the fineness is 10^{-4} denier. Further properties of nanofibers made by electrospinning will be discussed later in some detail.

2.3 Theoretical Analysis

2.3.1 Droplet Deformation and Jet Initiation

The transition of a bulk fluid material – either a melt or a solution – into extremely fine fibers, i.e., nanofibers with diameters down to a few nm, possibly with unique morphological and topological features, involves a sequence of complex deformation processes. These depend, on the one hand, on external parameters characteristic of the electrospinning technique itself (such as the applied field, the electrode configurations, or the feeding rate of the fluid to be spun) and, on the other hand, on intrinsic parameters characteristic of the spinning fluid itself (such as the surface free energy, the electric conductivity, and the viscous and elastic properties).

2.3.1.1 Droplet Deformation

The primary step in nanofiber formation involves the initiation of a fluid jet emanating from the spinning fluid due to its interaction with the electric field. This jet will in general not be initiated on flat fluid surfaces but rather at protrusions. Such protrusions can have the shapes, as previously discussed, of pending droplets located at the tip of syringe-like dies or of sessile droplets positioned on flat solid surfaces. They may also be formed via metal spikes arranged along metal wires or along metal cylinders encapsulated by the spinning fluid or even through statistical surface modulations induced by various means. Many more approaches along this line seem feasible.

The interactions of droplets/fluid protrusions with electric fields were considered in a set of papers going back as far as to the year 1882 concentrating predominantly on pendent and, to a lesser extent, on sessile droplets [27]. A frequent assumption is that the fluids to be spun such as polymer solutions or polymer melts tend to display an ionic conductivity to some degree. Within the electric field the anions or cations become nonuniformly distributed on the surface of the droplets in such a way that the surface becomes equipotential and the field inside the droplet zero. The general result of the theoretical analysis is that such droplets are deformed within the electrical field and display a stable critical shape even when close to a critical electric field beyond which jet formation occurs from the tip of the deformed droplet (Fig. 3). The shape should in principle be controlled by the equilibrium between the electric forces and surface tension as far as viscous and viscoelastic fluids are concerned, whereas nonrelaxing elastic forces may also affect the shape of the droplet. Yet, the papers on droplet deformation differ in the mathematical approaches taken and in their predictions on the shapes of the deformed droplets.

The interaction of fluid droplets with electric fields is a topic which was met with interest as long as 100 years ago. One motivation was the belief that the disintegration of water droplets in strong electric fields plays a major role in the formation of thunderstorms. In 1882 Rayleigh calculated the limited charge an isolated droplet can carry before it becomes unstable [27]. Using the same approach Zeleny analyzed the deformation of a droplet in an external electric field representing the shape of the droplet by a steroid [28]. Rayleigh subsequently argued that this approach is unsound for a droplet in an external electric field and calculated the deformation of the droplet due to the balance between internal pressure, surface tension, and electric forces (assuming also a spheroid as the representative shape of a water droplet) [29]. One problem of the analysis is the field dependence of the shape of the charged droplet and vice versa so that approximations have to be introduced in the analysis of the balance of forces. One approach, for instance, is concerned with the balance just at the poles and the equator of the deformed droplet. To be able to consider not only droplet deformation but also the mechanics of jet formation additional assumptions had to be introduced, such as a power law scaling of the electric potential [13, 29]. Predictions of the treatment by Tailor are that the droplet assumes a prolate shape and becomes unstable if the ratio of the length to the equatorial diameter approaches 1.9. Furthermore, the droplet is predicted to approach a conical shape close to the critical field with a half angle of the cone of 49.3° at the tip of the droplet. One finds half angles significantly smaller than this value in experiment (Fig. 3).

The problem of the shape of droplets in electrical fields was therefore revisited in the context of electrospinning (in particular by Yarin et al. [20]). First, they pointed out the self-similar nature of the treatment by Taylor. Second, they stated that the likely shape of a droplet in electrospinning must be very close to a hyperboloid of revolution. They then looked for self-similar and non-self-similar solutions for the balance of electric and mechanical forces and the corresponding shapes for hyperboloidic liquid droplets. The self-similar approach turned out to fail whereas the non-self-similar approach yielded results in agreement with the experimental ones. Based on the hyperboloidic approach, they predicted that the stationary critical shape assumed by the droplet in the critical electric field can be represented by hyperboloids approaching a conical asymptotic with a half angle of 33.5°. For sessile droplets the experimental analysis yielded values for the half angle that were close to the tip of the droplet of about 37° decreasing to 30.5° at locations farther away from the tip. In fact, the half angle experimentally observed for pending droplets close to the tip amounts to 31° decreasing to 26° at locations on the droplet surface further removed from the tip. In all cases these values are closer to the ones predicted by the hyperboloidal approach as compared to the spheroidal approach. A further interesting result of this analysis was that the curvature of the deformed drop at its tip is very small, amounting to about 600 nm. It can thus not easily be analyzed by optical means.

2.3.1.2
Jet Deformation

A highly important step in electrospinning that controls the fiber diameter as well as the orientational order and mechanical properties of the resulting nanofibers via self-assembly steps is the onset and further evolution of the bending instabilities. These were consequently analyzed in some detail by a set of theoretical approaches. It was pointed out by Yarin et al. that this feature in electrospinning and possibly other related phenomena seem to follow a very general theorem of electrostatics formulated by Earnshaw [18, 19]. According to this theorem it is impossible to prepare stable structures such as fluid jets in which all elements interact only by Coulomb forces. Charges located within the fluid jet, in our case, move the polymer material to which they are attached along complex pathways in such a way that the Coulomb interaction energy is minimized.

To be more specific even a simple linear arrangement of three equal charges arranged along a chain becomes unstable towards lateral deflections. Following this line of arguments Yarin et al. modeled the bending instabilities and the jet path resulting from them by a system of connected viscoelastic dumbbells with the beads in the dumbbells having masses and electrical charges representative of the charged polymer jet to be modeled [19]. The beads, in turn, interact among each other via springs and dashpots representing the mechanical interactions in the jet as well as its viscoelastic response towards the elongation during electrospinnning. The beads furthermore interact with each other by Coulomb forces since they carry charges. Finally, the interactions between the charged beads and the applied electric field are taken into account. Based on this straightforward model the evolution of the path of the fluid jet in the presence of bending perturbations was calculated. It turns out that the model provides a quite accurate description of the path of the jet, including the complex looping and spiraling motions (Fig. 4). This holds, for instance, not only for the conic envelop of the path within the bending instability range, but also for the magnitudes of the diameter reduction, total elongation, and deformation rate to which the jet is subjected.

This model, however, does not take into account the evaporation of the solvent during fiber formation or the onset of solidifications induced by the evaporation and the glass formation or crystallization, respectively. In fact, these processes affect the path of the jet to a considerable extent. The conic envelope representing the looping and spiraling of the jet in its bending mode is strongly extended both in the lateral and longitudinal directions as compared to the case neglecting evaporation and solidification. Further, the magnitudes of the total elongation and rate of elongation are affected as already discussed above [30]. The authors point out that details of the evaporation process and solidification process could not be taken into account since these details depend on the system studied, are actually unknown for most experimental systems, and have to be analyzed for the specific system of interest. Thus, the theory is able to give general estimates that definitely point into the right directions, as made obvious from a comparison between the experimentally observed features and the calculated ones. An interesting point of the theoretical analysis is that the formation of bending instabilities becomes suppressed if the surface energy of the jet becomes sufficiently large, keeping all other spinning parameters constant [13].

The observation that electrospinning does not always give rise to fibers with uniform diameter, but rather – for specific operating parameters – to fibers displaying modulations of the diameter, droplets or spikes arranged along the fiber length, to the deposition simultaneously of fibers and droplets on the counter electrode, or even just of droplets (Fig. 7a,b) are indications that the electrospinning process is even more complex than discussed so far. Observations show that frequently droplets form along the length of the jet in its straight part and that the slender jet/droplet arrangement subsequently becomes subjected to bending motions. On the positive side, these observations point out in which way, by a suitable choice of spinning parameters, the range of nanofiber architectures can be considerably expanded, which might be beneficial for specific applications.

Through their theoretical approaches, Hohman et al. have investigated the richness of structure formation processes taking place in electrospinning going beyond the bending instability (such as the axisymmetric instability depicted in Fig. 8 in addition to the bending instability) [31–34].

In fact, they were able to predict first of all phase diagrams specifying for which spinning parameters which kind of structure formation process/type of instability is dominant, and they secondly specified operating diagrams for electrospinning in terms of the feeding rate of the spinning solution and the applied electrical field for given properties of the spinning fluids. Such diagrams (to be discussed in more detail below) allow selection of electrospinning parameter sets in such a way that the bending mode becomes dominant yielding smooth fibers with homogeneous fiber diameters. Yet, they also allow selection of these parameters in such a way that other types of instabilities take over yielding, e.g., fibers with droplets attached to them, thus corresponding to the electrospraying mode.

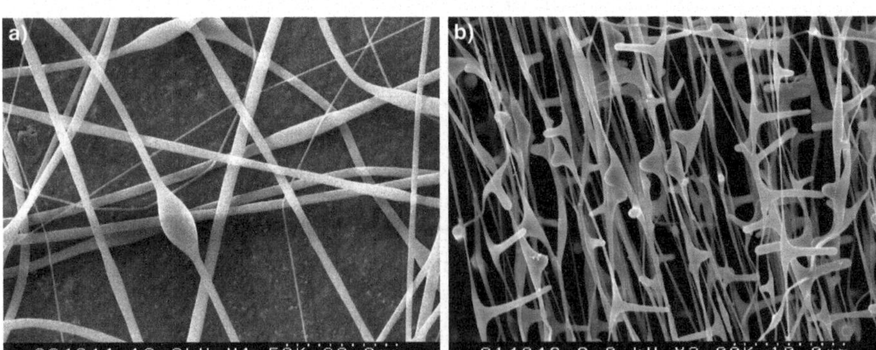

Fig. 7 Electrospun nanofibers: **a** With beads (PS) and **b** With spikes (barbed nanowires) along the fibers (PVA)

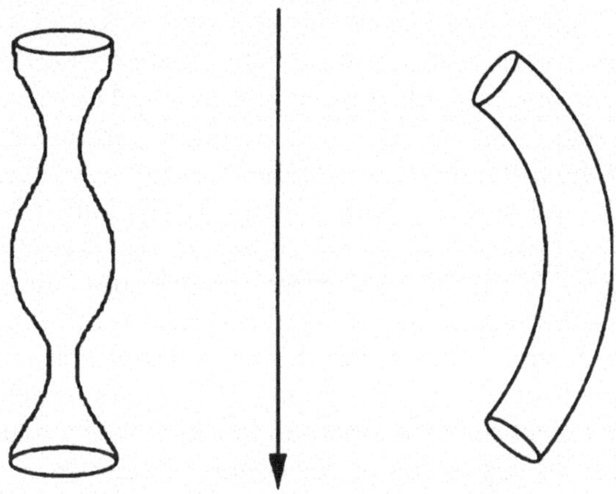

Fig. 8 Schematic sketch of an axisymmetric instability (*left*) and bending instability adapted from Shin et al. [33]

Hohman et al. based their theoretical analysis on classical hydrodynamics adapted to the particular case of long fluid cylindrical elements carrying charges and being located in an electric field – mimicking the jet in electrospinning [31–34]. Details of the theoretical analysis will not be spelled out in this contribution but should be explored in the respective papers. This contribution will rather concentrate on major predictions of interest for the experimentalist. In this context it is sufficient to point out that the treatment has to consider in addition to the conservation laws for the mass and the charges the presence of viscous dissipation, of electric forces arising from the coupling of the charged fluid elements to the electric field but also of gravitational forces. Electrostatic terms were thus introduced into the hydrodynamic equation. Furthermore, to facilitate the theoretical treatment and to adapt it to the situation of a long, slender fluid cylinder with a specific axial ratio expression for the flow velocity and electric field components (radial and tangential) were expanded in a Taylor series in powers of the aspect ratio of the slender fluid cylinder and introduced into the electro-hydrodynamic equations.

The theoretical analysis comprises two different steps:

(i) In the first step, the stability of this fluid electrically charged element is investigated at fixed charge density, surface tension, viscosity, and dielectric constant.

(ii) In a second step the treatment takes into account the fact that these properties vary along the jet as it moves from the die to the counter electrode.

The analysis first of all specifies the types of instabilities to which such a fluid element is subjected.

The first one is the electric counterpart of the classical Rayleigh instability known from uncharged fluid threads, already discussed above. This type of instability is controlled by surface energy contributions, and it consists in the growth of diameter perturbations eventually causing a breakup of the long cylindrical element into individual isolated droplets. The presence of surface charges tends to reduce the effect of the controlling parameter surface energy on the growth of the instability. The instability becomes weaker with increasing electric fields and surface charges, respectively, and it becomes totally suppressed above a critical field for which the electric pressure coming from the surface charges exceeds the surfaces tension pressure. The critical field depends linearly on the surface tension and inversely on the radius of the jet. Furthermore, the analysis shows that this type of instability does not cause a breakup of the jet for the set of spinning parameters usually used in electrospinning. One exception is the situation in which the jet diameter becomes very small, close to its deposition on the counter electrode.

The second type of instability observed (also for slender fluid elements) causing a growth of diameter perturbations with a final breakup is totally controlled by charge contributions rather than by surface energy. In the case of the charge driven axisymmetrical instability, a statistic variance of the jet's radius causes a modulation of the surface charge density. This in turn generates tangential electrical forces that couple to the radius modulation and amplify it. The formation of beads is the end result of such a coupling loop. In fact, during electrospinning fibers on which drops along the fiber are aligned like pearls on a string can be observed for certain sets of spinning parameters (Figs. 7 and 8).

Finally, the theoretical analysis by Hohman et al. [31–34] yields not unexpectedly the bending instability discussed already in detail previously and called whipping instability in these papers. An important outcome of this theoretical analysis for experimentalists is the prediction of phase diagrams for the various instabilities and of operating diagrams specifying for which set of operating parameters electrospinning can be performed to produce the bending instability yielding homogeneous fibers respectively: the whipping or bending mode is enhanced if the local electrical field near the jet is dominated by its own charges, and it is suppressed if the local field is governed by the external tangential field (Fig. 9).

Figure 9 displays a characteristic operating diagram specifying the dominance of either the bending or the conductive axisymmetric instability as a function of the operating parameters flow rate of the spinning solution and applied field for given values of fluid parameters (such as the surface tension, viscosity, electric conductivity, and dielectric constant).

The general trends of such operating diagrams do not change significantly as the fluid properties are modified but rather just the absolute numbers vary. The operating diagram reveals which way the field and flow rate have to be controlled to yield the bending instability and thus ensure stable electrospinning of smooth nanofibers. In experiments, one has to determine the fluid parameters mentioned above and then one should be able to optimize elec-

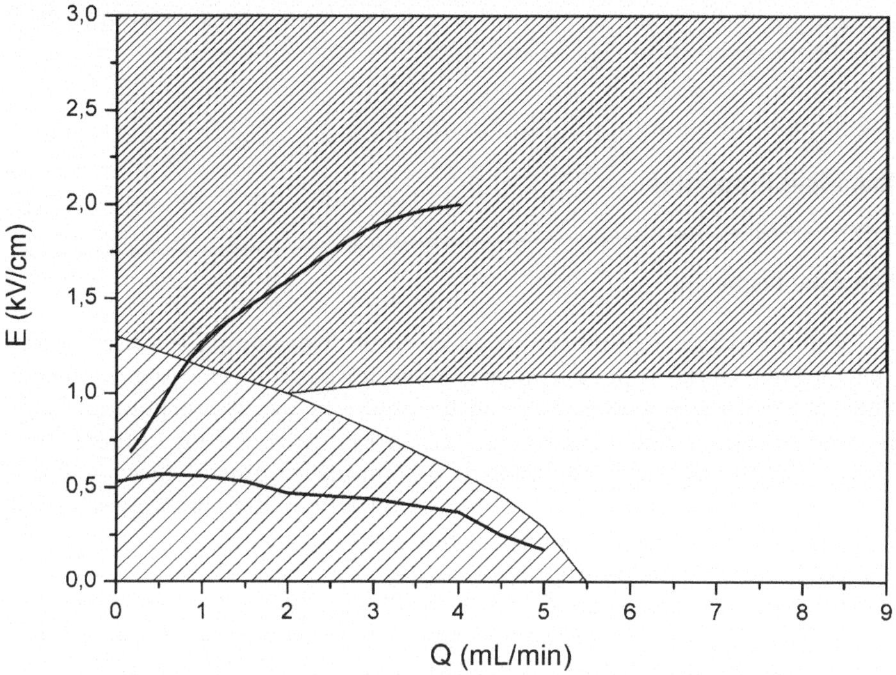

Fig. 9 Example of an operating electric field versus feeding rate diagram for electrospinning adapted from Hohman et al. [32, 33]. The *upper shaded area* shows the theoretically predicted onset of bending instabilities; the *lower* one shows the corresponding onset of axisymmetric instability. The *lines* represent experimental results on the instability thresholds for the two types of instabilities for PEO solutions for a given set of electric conductivity, viscosity, dielectric constant, and surface free energy values corresponding to the ones assumed in the theoretical treatment

trospinning by a suitable choice of field and flow rate on an absolute scale. Further, following the theoretical treatments discussed above, Fridrikh et al. presented a simple analytical model in terms of current and feeding rate, allowing prediction of the terminal jet diameter (beyond which a further thinning due to bending instabilities no longer occurs, as the stresses from surface tension and from surface charge are in balance) [35].

Finally, as far as investigating the electrospinning processes is concerned, it should be pointed out that more phenomena may be observed during electrospinning. One is the occurrence of jet branching characterized by periodic arrangements of the branches extending perpendicular to the jet [13]. Yarin et al. developed a electrohydrodynamic theory which shows that the surface of conducting fluid jets can display complex static undulations at strong electric fields [36]. Such undulations can become unstable for specific conditions leading to the ejection of regular arrangements of lateral branches from the primary jet. It seems feasible that various types of fiber structures may result such as fluffy columnar networks, called garlands fibers, or even barbed nanofibers as shown in Figs. 1f and 7b.

3
Nanofiber Properties

3.1
Nanofiber Diameters

The diameter of the nanofibers produced by electrospinning is a key parameter for most of the applications envisioned for such fibers. Fiber diameters down to just a few nanometers can be produced by electrospinning from polymer solutions but the diameter can also be extended up to more than 10 μm if required. The most influential spinning parameter to control the fiber diameter is the polymer concentration within the spinning solution. A lower polymer concentration will cause first of all a stronger fiber diameter shrinkage solely by solvent evaporation. Yet, the effect of polymer concentration goes well above this direct contribution, since, in general, polymer concentrations suitable for spinning can be varied only by a factor of well below 10, though fiber diameter variations amounting to a factor of 100 and more are known. A major effect of a variation of the polymer concentration on resulting fiber diameters comes from the corresponding strong variation of the viscosity and the viscoelastic response of the fluid jet to deformations. Both the viscosity and viscoelasticity of polymer solution represented, e.g., by the corresponding relaxation time, are known to strongly depend on the polymer concentration. As far as electrospinning is concerned the variation of these parameters in elongational deformations involving high deformation speeds has to be taken into account. Studies on the capillary thinning process of threads of dilute and semidilute polymer solutions revealed that the deformation happens in two stages the first involving the viscoelastic stretching of polymer coils and the second a quasi-Newtonian flow in which fully stretched coils flow past each other giving rise to a constant elongational viscosity [37]. A further finding which may also be of importance for electrospinning is that such strong and rapid elongational deformation tends to induce chain rupture to a considerable extent.

The dependence of the fiber diameter on the polymer concentration has been investigated for a set of polymers including among others polyamides, polylactides, polyacrylonitrile, polyvinyl acetate or polyethylene oxide [9–13]. One characteristic example is revealed in Fig. 10 for PA 6 dissolved in acetic acid [38].

The diameter changes for polyamide 6 by a factor of more than 10 covering the range from about 150 nm up to about 1.7 μm keeping all other spinning parameters, such as the applied voltage, electrode distances con-

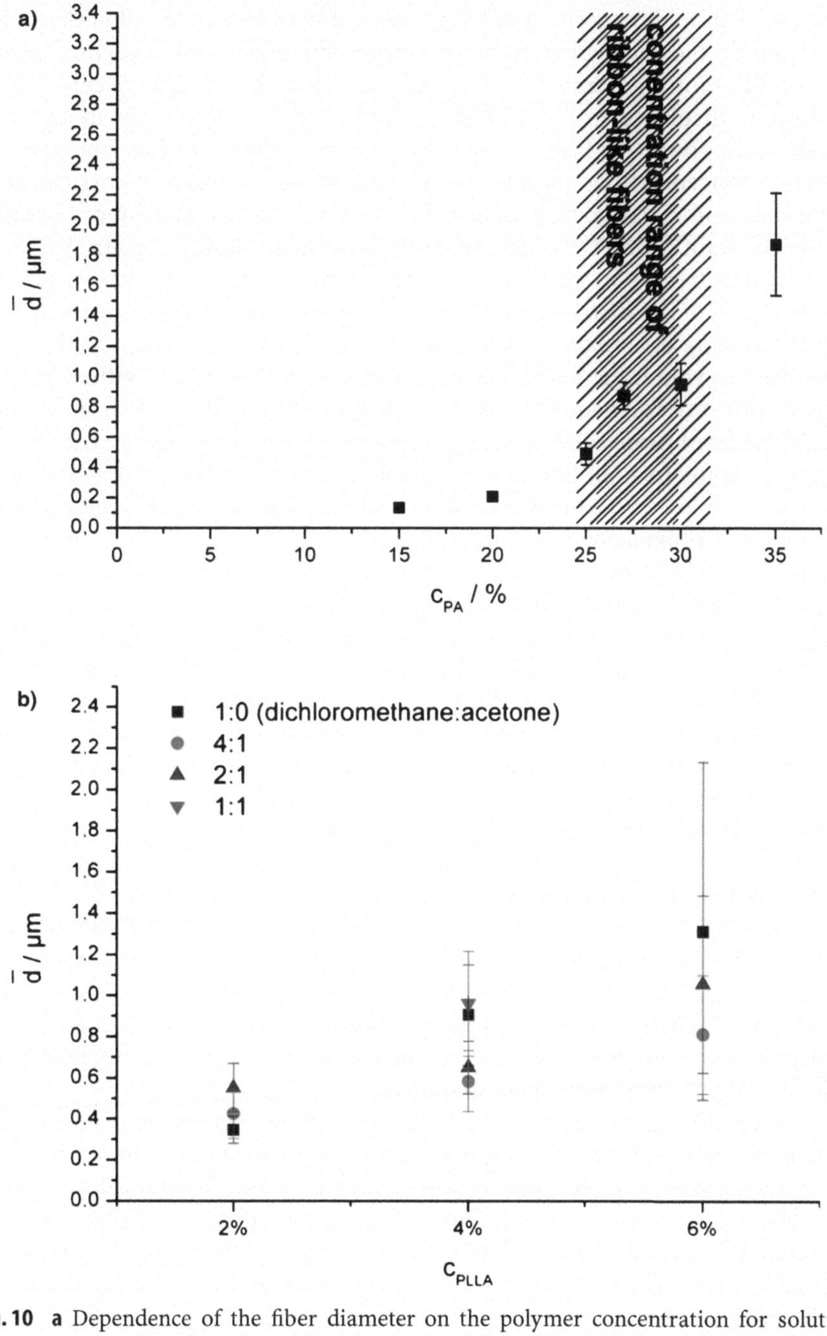

Fig. 10 **a** Dependence of the fiber diameter on the polymer concentration for solution electrospinning [38] for polyamide 6 in acetic acid; in the *shaded area* band-structures rather than fibers with circular cross sections are obtained, **b** Corresponding variation for polylactide in different solvent mixtures

stant, etc. It should be pointed out that one is able to reduce the fiber diameter of polyamide 6 even further to below 50 nm, yet this involves further variations of the spinning parameters beyond the polymer solution concentration. The finding displayed in Fig. 10 a is that the fiber diameter tends to increase in a strongly nonlinear fashion with the polymer concentration, particularly at higher concentrations (similar to the case of corresponding variations of the viscous and viscoelastic properties). The fiber diameters achieved are not strictly constant for a given concentration but rather show a certain distribution which increases in absolute numbers as the diameter increases, yet the relative width of the distribution frequently tends to remain about constant. It is an interesting observation (also indicated in Fig. 10a) that a concentration range exist in which flat band-shaped fibers rather than fibers with spherical cross section are formed. This aspect will be considered below in more detail. Polymer solutions with concentrations below the range showed in Fig. 10 cannot be spun to nanofibers since their tendency to droplet formation is strong at such low concentration unless specific additives modifying the conductivity and other properties are added. On the other hand, solutions with polymer concentrations above the upper values shown in Fig. 10 are, in general, so viscous that again electrospinning fails.

Figure 10b shows corresponding polymer concentration-fiber diameter results for polylactide in dichloromethane. It is obvious that suitable variations of the polymer concentration and of the resulting fiber diameters are much smaller than found for polyamide and that the distribution of the fiber diameters at a given concentration is also broader. These results reflect the different nature of the polylactide solutions in terms of molecular weight, molecular weight distribution, entanglement, interaction with the solvent (which affects both the viscous and viscoelastic properties of the solution). In fact, detailed studies are still sparse trying to relate fiber properties to real data on viscous and viscoelastic properties. A major reason is, of course, that fiber spinning is performed with extremely high strains and strain rates and that, in general, very limited data are available for the viscous/viscoelastic properties for such extreme conditions. Thus, for the time being one has to investigate fiber diameter-polymer solution correlations experimentally for each case of polymer/solvent combinations.

An interesting approach to modify the viscous at viscoelastic properties of a given polymer solution during electrospinning involves the application of ultrasonic agitation to the solution just as it leaves the capillary die [39, 40]. Two effects have been reported. One is that electrospinning of fibers can be extended to higher polymer solution concentration well outside the range accessible without ultrasonic agitation and the other one is that fiber diameter can be modified to a certain extent in this way.

A second spinning parameter which can be exploited to adjust the fiber diameter is the feeding rate, i.e., the amount of polymer solution that is fed per time unit into the spinning apparatus at constant die diameter, and keeping all other spinning parameters constant. Among the examples to be found in the literature are studies on polyacrylonitrile (PAN) which was studied both experimentally and from a theoretical point of view [41]. The result is that again the fiber diameter can be varied in this case by more than a factor of 10 by controlling the feeding rate. Similar results have been reported for other polymer solutions yet it seems that the magnitude of fiber diameter modifications induced via the feeding rate depends strongly on the polymer system under investigations and frequently tends to be rather small.

Polymer concentration and polymer solution feeding rate are not the only spinning parameters that have been evaluated with respect to their impact on fiber diameters in electrospinning. The jet diameters and the fiber diameters can be controlled to a certain extent also by the applied voltage, though again the achieved results vary strongly with the polymer system. It was reported, for instance, for electrospinning of acrylic nanofibers that the jet diameter may decrease initially with increasing voltage, yet that it increased again as the voltage was further increased [42, 43]. This is obviously due to a strong increase of the amount of polymer solution drawn out by the electric field from the capillary. A further approach towards controlling the fiber diameter consists in extending the range of polymer concentrations suitable for spinning to much smaller concentrations and thus smaller polymer fiber diameters by adding components which vary the electric conductivity of the solvent [44]. An increased conductivity tends to increase the charge density at the surface of the jets, thus decreasing the tendency of droplet formation during electrospinning. Finally, a variation of the distance between the spinning die and the counter electrode has been used to affect the fiber diameter. Various effects may contribute to fiber diameter variations, e.g., as the distance is decreased among them an increase in the electric field, a suppression of later stages of elongational processes induced by the bending instability, or the suppression of the complete evaporation of the solvent may have effect. It is for this reason that again the induced fiber diameter variations vary strongly with the polymer system under consideration. Thus, the general conclusion is that as far as solution spinning is concerned the concentration of the polymer solution is the most direct and most general parameter to control fiber diameters in electrospinning.

3.2 Shape of the Fibers

The goal of electrospinning will be in the majority of cases the production of nanofibers uniform in diameter along their complete length and with a spherical cross-sectional shape. In fact, electrospinning tends to produce just such a type of fibers in the range of polymer solution concentrations in which droplet formation is suppressed. This is not really surprising since the geometries of the dies, of the applied field, and, thus, of the deformation of the droplets are in the majority of cases axially symmetric. Furthermore, the fibers tend to be straight for a significant persistence length often larger than 10 cm and above. This becomes particularly evident if the fibers are deposited in a parallel fashion due to particular electrode configurations to be discussed below in more detail. Yet, in several cases fiber shapes have been reported showing strong buckling (Fig. 11) [19]. Buckling obviously results from the presence of longitudinal compressive forces acting on the impinging thread. Detailed investigations of the buckling phenomenon have revealed a surprising richness of buckling pattern including sinusoidal trajectories, meandering, coiled structures, figure eight structures double pattern, and many more. Further, the trajectories of the deposited nanofibers may result from a superposition of the looping motions due to bending instabilities and buckling.

Nanofibers with spherical cross sections are the general target for the majority of applications introduced previously. Yet, considering the prism-shaped silk fibers and the corresponding peculiar optical properties, other types of fiber cross sections may be beneficial for specific applications. In fact, electrospinning yields, either by poor controlling or on purpose by an appropriate controlling, a variety of fiber cross-sectional shapes originating from the complex self-assembly processes intrinsic in electrospinning.

One frequent observation is the formation of band-shaped fibers characterized by a flat rectangular cross-sectional area [45]. An example of such a fiber shape is shown in Fig. 1b. It results, for instance, in a very limited range of polymer concentrations in the case of polyamide 6 spun from acetic acid solutions, while fibers with spherical cross-sectional shape are formed outside this concentration range. Band-shaped fibers have also been reported for polymers such as polycarbonate and many more.

A tentative explanation for the formation of such band-shaped fibers in electrospinning is that the solvent evaporates particularly rapidly in many in-

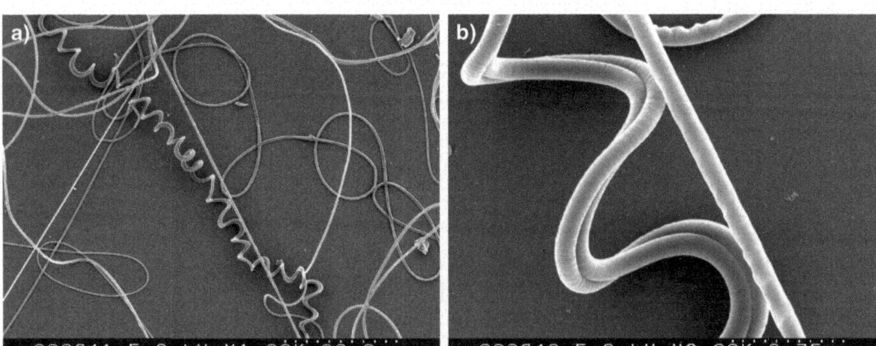

Fig. 11 Buckling PA 6 nanofibers electrospun from formic acid solutions. (Fibers shown for different magnifications)

stances from the surface of the fluid jet giving rise to a solid surface shell [45]. The subsequent evaporation of the remaining solvent in the core area gives rise to hollow fibers. These in turn are assumed to collapse in such a way that band-shaped fibers result. In fact there are indications based on optical studies that such a process actually takes place in certain cases. Yet, the observation for polyamide 6 systems that the formation of band-shaped fibers is restricted to a very narrow concentration interval with fibers displaying a spherical cross section forming on both sides of this concentration interval suggests that this may not be the only mechanism yielding such special fiber shapes [38]. In any case such fiber cross-sectional shapes can be prepared reproducibly and they could well be of interest for specific applications. Nonwovens composed of them should display total porosities, pore sizes, and permeation properties differing significantly from those composed of fibers with spherical cross-sectional areas.

Fibers characterized by droplets or spindle type elements arranged along the fiber in a regular or random fashion is a further frequently observed nanofiber structure in electrospinning. Figures 1d and 12a give examples of such structures. The origin of such structures seems to be that the bending instability tending to produce nanofibers with uniform diameter becomes superimposed by other instabilities notably the axisymmetric instability as suggested by the theoretical operating diagram discussed previously. Spinning nanofibers from dilute solutions, from solutions in which the electric conductivity is below a critical value, and also spinning polymers for which the molecular weight is too small (meaning no chain entanglements are formed) are reasons for such fiber structures. These are in the majority of cases unwanted and can be suppressed by a suitable variation of the processing parameters. Adding specific low molar mass compounds to enhance the ionic conductivity is one possible approach that has been demonstrated for various polymer systems [44]. On the other hand in special cases like drug delivery such structures may be beneficial and can thus be produced reproducibly.

Another feature of interest is fragmentation of electrospun nanofibers to linear segments or droplets. In general the integrity of the nanofibers is a major requirement for many of the applications introduced above among them textiles, filter applications and in particular also nanofiber reinforcement. This integrity should survive processing steps involving heat treatment occurring during filter formation or textile production, or during the incorporation of the fibers into a polymer matrix to be reinforced. Furthermore, applications may also require the nanofibers to be stable at enhanced temperatures for longer time intervals. Nanofiber modified surfaces displaying ultrahydrophobic properties [46] might also be subjected to further coating steps involving again enhanced temperatures. The integrity of the fibers is a requirement in all cases, i.e., fragmentation processes should be absent.

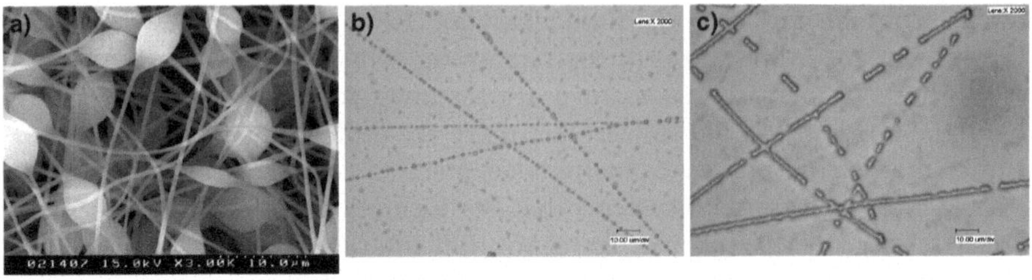

Fig. 12 **a** Beaded fibers (PS), **b** PA 6 nanofibers decomposing into droplets via Rayleigh instabilities. **c** Polystyrene nanofibers decomposing to linear segments as intermediate states of fragmentation [89]

In other cases fragmentation might be something which is wanted. A complete or partial fragmentation of nanofibers to nanodroplets may be of interest for further modification of surface properties of substrates involving superhydrophobicity [46]. To induce such structures in a highly controlled way one may also start from solid nanofibers and anneal them at elevated temperatures where the polymers become soft. Conventional Rayleigh instabilities controlled by the surface energy can thus be induced, which cause the formation of a pearl necklace structure with a uniform size and spacing of the droplets along the fiber, as obvious from Fig. 12b [47,48,89]. The diameter of the droplets and their distance is controlled to a major extent in this case by the original fiber diameter. As intermediate state, a fragmentation into linear segments that can be conserved by cooling may appear as shown in Fig. 12c. Self-organization may thus be used to produce nanowires from nanofibers.

Finally, complex fiber structure with branchings, spikes extending from the fiber backbone, have been observed (see examples in Figs. 1f and 7b). Up to now no explanations exist for their formation other than that they may be the consequence of a further pathway following self-organization controlled by Coulomb interactions possibly involving splaying at its early stage. Again such structures can be produced in a reproducible manner and may find specific applications in the future.

3.3 Nanofiber Topologies, Porous Fibers

Electrospinning generally yields fibers with smooth surfaces, as shown in the discussions above, if the key electrospinning parameters are controlled in such a way that the self-assembly processes are directed along this line. Yet, keeping these spinning parameters constant but choosing the spinning solution appropriate to other types of self-assembly effects can be superimposed. Phase separation between solvents and polymer species both for the case of a single solvent or solvent mixtures, phase separation effects between polymers in spinning solutions composed of more than one polymer, between a polymer and low molar mass additives, and condensation effects of water vapor initiated on the surface of the jet can be exploited in a controlled way to give rise not only to specific supermolecular structures but also to specific surface topologies.

It is well known from thermodynamics of mixtures that phase separations may take place leading to two types of phase structures depending on the concentration ranges considered, the molecular interactions between the components, the temperature, on the time scale in which separation takes place and also on the mechanical stresses present. It is rather difficult to make quantitative predictions for the electrospinning process even if the phase diagrams are known for the solutions under study. Qualitatively two types of phase morphologies, the bimodal or the spinodal type, should occur if phase separation takes place during electrospinning respectively [49]. Phase separated domains dispersed within a continuous matrix are characteristic for the bimodal case and a co-continuous phase morphology with a certain regularity of the arrangements of the different phases for the spinodal case. In electrospinning of fibers the final phase morphology found in the fibers and the composition of the phase are modified by the evaporation of the solvent and the strong elongational deformation.

One direction which has been investigated along this line is preparation of porous fibers with various types of pore sizes, pore arrangements as well as with various magnitudes of the total porosity and thus of specific surface areas. These are of interest for a variety of applications, e.g., tissue engineering, filter technique, catalysis, drug delivery, nanofiber reinforcement, and many others. Pores can function as anchorage for cells in tissue engineering. They may increase the surface in filter applications or catalysis by up to one order of magnitude, may modify wetting properties and thus matrix fiber coupling in the case of fiber reinforcement, and can be exploited to modify the release kinetics of drugs. Pores can also influence the kinetics of biodegradation of biodegradable nanofibers.

The most direct way to produce porous fibers consists in spinning the fibers still containing solvents on very cold surfaces or directly into liquid nitrogen causing in both cases an instant phase separation into solvent and solid polymer [50]. Freeze-drying approaches can then be used to remove the solvents from within the solid polymer fibers yielding highly porous polymers. This was shown for polystyrene, polyacrylonitrile, polycaprolactone, as well as polyvinylidenfluoride. One disadvantage of this approach is that one needs a subsequent treatment of the fibers (freeze drying) and a transfer of the fibers to the target substrate

A second and somewhat more direct approach is based on water vapor condensation processes in humid environments [51, 52]. The model consideration is that the spinning jet is cooled down due to solvent evaporation and that the tiniest droplets of water consequently precipitate onto the spinning jet. These droplets then form the pores in the solidified fibers. The extent of pore formation and also pore sizes can be tuned by variation of humidity. The pores which result tend to assemble predominantly on the surface of the fibers and to display a spherical shape. It thus seems that they frequently do not experience longitudinal deformations taking place in electrospinning to an appreciable extent.

Another approach towards porous structures relies on phase separation into polymer rich and polymer depleted regions as evaporation of the solvent takes place (this can generally be roughly estimated from phase diagrams), yielding polymer rich regions and polymer depleted regions that will form pores in the solid phase if the solvent is completely evaporated [53]. The extent of the pore formation is thereby defined by the relative percentages of both phases. Employment of solvent mixtures allows a selective adjustment of porosities. Figure 1c illustrates examples of porous polylactide nanofibers that were fabricated by this method. It is apparent that the degree of porosity can be varied strongly and that the pores may be elongated.

Entirely different fiber topologies are obtained if nanofibers are spun from a solution of mixtures of immiscible polymers in the same solvent [53]. Phase separation occurs with the evaporation of the solvent, leading to binodal or spinodal structures, respectively, in the nanofibers depending on the system. To generate porous fibers of high overall porosity two strategies are possible [54]:

1. Highly swelling solvents induce significant increases in fiber diameter and a highly porous structure (Fig. 13) with porosities of 75% and more remains after removal of the swelling agent.
2. Removal of one of the phases in the case of the previously mentioned binodal or spinodal segregation by specific solvents yields highly porous

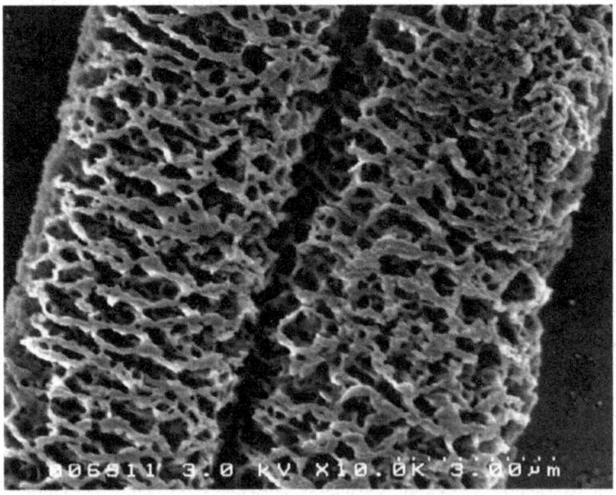

Fig. 13 Porous nanofibers obtained via phase separation processes during electrospinning for mixtures of polylactide and polyethylene oxide [54]

fibers even with periodic arrangements if a spinodal phase separation between the two polymers under consideration has taken place.

Approaches based on phase separation of polymer components and the selective removal of one of the components may also be directed in such a way that fibers without pores yet characterized by fractal surfaces (Fig. 1e) result. Such types of fibers can be predominantly obtained for specific concentration ranges of the two polymers and for spinodal decompositions.

3.4 Internal Morphology

The internal structure of electrospun nanofibers will be controlled in principle by three major contributions:

1. Glass formation and crystallization as in the bulk, governed by thermodynamic and kinetic factors;
2. Rapid mechanical deformation characteristic of the jet deformation in electrospinning;
3. Confinement effects arising from the small scale of the resulting fiber diameters.

In the majority of cases nanofibers are produced by electrospinning from solutions. Fiber formation is controlled by the simultaneously proceeding processes of solvent evaporation and strong elongation of the solidifying fibers. A volume element of the jet travels the whole way from the spinning die to the final deposition of the solid fiber on the counter electrode in a time frame of typically 10^{-1} s. The time span for the structure formation within the resulting nanofibers is therefore typically 10^{-2} s and below. In this respect, electrospinning resembles spincoating in which nanofilms are cast from solution by transferring droplets of the solution onto a rapidly rotating substrate. Freezing-in of amorphous regions into a glassy state, a partial crystallization incorporating the formation of lamellas, the formation of orientational orders for the chain molecules and the crystals, and the formation of supermolecular structures (such as spherulitic structures or phase domains as far as blends are concerned) (see above) are processes happening in both electrospinning and spincoating within the solidifying fibers or films during a very short time scale. One major topic to be discussed in the following is the effect of the short time scale on the structure formation and on the properties of the resulting structures. The second major topic is the effect of the confinement imposed by the small diameter of the fibers on structure formation and properties. It is well known that confinement affects structure formations, molecular dynamics and properties [55]. In the following the discussion will distinguish between polymers unable to crystallize, e.g., atactic polystyrene, atactic polymethylmethacrylate, and those that can, e.g., polyamides, polylactides, polyethylene oxide and many more.

3.4.1 Amorphous Polymers

A polymer that is unable to crystallize will experience a freezing-in process in a specific temperature range, which depends on the individual polymer that is being considered [56]. The freezing-in process does not correspond to a thermodynamic phase transition but is rather a kinetic transition from the supercooled molten state (a thermodynamically equilibrium state) to the nonequilibrium state of a frozen melt. This transition is accompanied by a rapid increase in relaxations times, viscosities, mechanical moduli. A characteristic feature of the glass transition is that the location of the transition temperature depends on the thermal history. It increases as the speed with which the cooling process proceeds is increased. Yet, despite the rapid fiber formation and solidification process characteristic of electrospinning no significant deviation of the glass transition temperature of nanofibers from the one of the bulk material has been reported so far.

The glass transition temperature has furthermore been claimed to depend strongly on confinement effects. A decrease of the glass transition temperature of several 10 K has been claimed for nanofilms as the film thickness approaches values of 10 nm and below. Yet, it seems that such strong shift can frequently be accounted for by degradation processes taking place during preparation and annealing of such thin films at elevated temperatures and that nanofilms prepared in vacuum or in nitrogen atmosphere do neither show degradation nor a significant depression of the glass transition temperature even for films with thicknesses in the 10 nm range [57, 58]. It is therefore not surprising that no depression of the glass transition temperature of nanofibers relative to the value of the bulk material has been reported so far.

A further feature of the freezing-in process is that not only the glass transition temperature but also properties within the glassy state, i.e., density, enthalpy, moduli, etc., depend on the speed with which the glassy state is approached [56]: the properties of the glassy polymers in nanofibers might strongly differ from those of fibers formed more slowly. One feature which can be looked upon as a manifestation of the particular state of the glass is the physical aging process. The nonequilibrium glassy state tends to relax towards the equilibrium state of the supercooled melt and this relaxation depends on the speed with which the glass was formed. Enthalpy relaxations related to aging and volume relaxations, again related to aging, can be analyzed by calorimetry and dilatometry respectively. Calorimetric studies reveal the glass transition in the absence of aging processes by a stepwise increase of the specific heat at the glass transition temperature on heating due to the onset of additional degrees of freedom. Aging processes of the enthalpy manifest themselves by the superposition of this stepwise increase by additional maxima or minima in the heat flow. There are only few statements concerning this phenomenon in electrospun nanofibers in the literature [59]. Electrospun nanofibers of semi-crystalline polylactide nanofibers show distinct deviation of the heat flow from the usual step-like increase for the amorphous regions, indicating aging processes

The induction of chain orientation via electrospinning in glassy polymers is a topic of considerable interest in basic research but also with respect to applications. The experimental analysis of electrospinning processes shows, as discussed above, that the elongation of the fibers is very high, up to a factor of 10^5, while the elongation rates simultaneously reach values of $10^5\ s^{-1}$. Estimations are made according to which chains become highly oriented, as in the case of when the product of the elongation rate and chain relaxation time is higher that 0.5 [25]. For polymer solutions suitable for electrospinning, the appropriate ranges of their molecular weights and polymer concentration relaxation times of 10^{-3} to 10^{-2} s were reported [13, 19, 37]. A significant chain orientation should thus be generated during electrospinning. It is still an unanswered question whether the deposited fibers continue to contain traces of solvent that would promote chain relaxation. This should depend on the vapor pressure of the solvent, amongst other parameters.

As a matter of fact the results of experimental studies on the occurrence of birefringence are rather inconsistent. In some cases a slight birefringence was observed, e.g., in polybenzimidazol nanofibers with diameters around 300 nm, as well as in nanofibers from styrene-butadiene-styrene triblock copolymers with diameters of about 100 nm, as shown by Kim et al. [60]. The latter systems, of course, already contain some kind of order. In other cases nanofibers that did not show birefringence and therefore obviously feature no chain orientation were obtained from purely amorphous polymers, such as polystyrene and others. It is not entirely understandable why amorphous nanofibers do not generally show high chain orientations in view of the deformation rates typical for electrospinning and of the relaxation times characteristic for the spinning solutions. There still quite clearly remains a need for systematic investigations. In any case it has to be anticipated that nanofibers displaying orientations will shrink if heated to higher temperatures or come into contact with solvents which cause swelling.

3.4.2
Partial Crystalline Nanofibers

An opinion which is frequently voiced as far as nanofibers formation by electrospinning is concerned is that the structure in nanofibers and the properties of the growing crystals should deviate strongly from the one of bulk systems due to rapid structure formation and confinement effects imposed by the small diameter of the fibers. Melting point depressions originating from a reduction in size of the crystals are well known from metals or semi-conductor systems. Yet, no such effects actually seem to be of general significance in electrospun polymer nanofibers. One has to take into account in this context that the crystallization in polymers, resulting crystal sizes and crystal shapes differ in many aspects from the one taking place in low molar mass materials [61].

One feature is that polymers tend to crystallize only to a certain extent characterized by a degree of crystallinity, which may amount to significantly less than 50%. The depression of crystallization originates primarily from the strong entanglements of the polymer chains in the melt or in highly concentrated solutions. These restrict configurational changes necessary for crystallization. A further feature is that polymers tend to form lamellar crystals with lamellar thicknesses in the few nanometer to 10 nm range already in the absence of confinement. The lamellar thickness and thus the melting temperature of the lamellae depend predominantly on the thermal history for kinetic reasons. Finally, the crystal structures usually contain defects, e.g., coming from chain ends. These features suggest the possibility that solidification of nanofibers comprising crystallization will not be too different from the one taking place in the bulk, which turns out to be true.

In fact, there are no explicit indications in the literature that solidification processes in electrospun nanofibers related to crystallization are significantly different from those in bulk. It was, for instance, reported for the case of nanofibers from PET, polyethylenenaphthalate, and mixtures of these polymers that the melting points are not modified to an appreciable extent by the processing of the polymers into nanofibers. At the same time, it was reported that the degree of crystallinity may be increased as well as the glass transition temperature and the crystallization temperature [62]. The authors, however, indicated that transesterification processes occurred in the polymer mixtures and that, therefore, a decrease of chain lengths of chain molecules during electrospinning has to be anticipated. In the case of polylactide, it was observed that the degree of crystallinity as well as the melting point of the electrospun fibers were very similar to those of macroscopically prepared samples [63].

There are several studies concerning electrospinning of polyamides. It was reported that electrospinning of PA6 yields nanofibers that form the less ordered γ-crystal modification as obvious from X-ray diffraction and Raman investigations [64, 65]. Crystallization from bulk solution resulted in the ordered α-form. In this regard it is interesting that in the case of conventional fibers fabricated by melt extrusion the formation of crystals proceeds similar to the one of the crystal formation during electrospinning, despite all differences in the fabrication of the fibers. This holds especially when the elongation after extrusion exceeds a definite limit. In this case the less ordered γ-phase is observed. Therefore, the reasoning is that the development of the γ-form in nanofibers is a distinct indication of strong mechanical deformations occurring during electrospinning. This phase can be converted into the higher ordered α-phase by annealing of the nanofibers at higher temperatures.

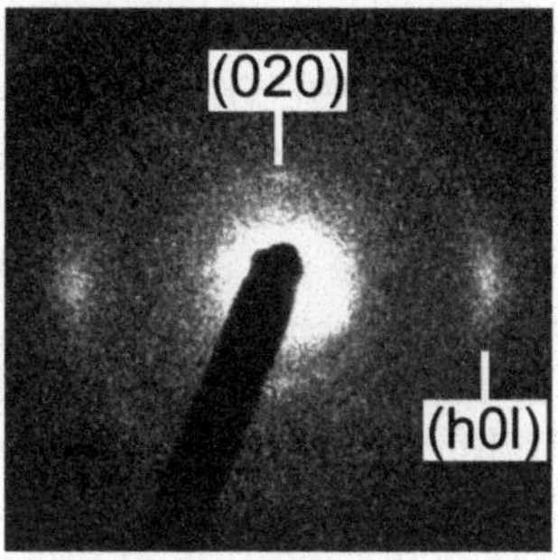

Fig. 14 Selected area electron diffraction diagram (SAED) obtained for a single PA 6 nanofiber [63, 65]

Of particular significance for mechanical properties of the fibers is the chain orientation and the orientation of the crystallites, which show up in the electrospun fibers as again made obvious from Raman investigations, X-ray or electron diffraction studies, and other indicators. Very high crystallite orientations in nanofibers from poly(ferrocenyldimethylsilan) were shown by Chen et al. who pointed out that the perfection of orientation could be substantially increased by annealing the fibers [66]. This orientational order is particularly striking if diffraction experiments are carried out on single fibers. Applying the so called SAED (selected area electron diffraction) technique the degree of orientation of the crystals in a PA6 nanofiber with a thickness of 50 nm was analyzed [63]. These diffraction experiments revealed the presence of very high crystallite orientations in the individual nanofibers produced by electrospinning (Fig. 14). This is apparent from the inhomogeneous azimuthal distribution of the diffraction intensity. Comparable orientations can only be achieved in macroscopic fibers produced by melt extrusion if the extrusion is followed by extreme deformations.

For the particular case of nanofibers made from liquid crystalline polymers, using a polyhexylisocyanate as example, it was demonstrated that the degree of orientation can depend greatly on the fiber thickness. Jaeger et al. furthermore report that in PEO fibers especially surface layers may be highly oriented [67]. Finally, polymer nanofibers composed of a single nematic monodomain were observed for main chain liquid crystalline elastomers, which were cross-linked during electrospinning via UV-light irradiation [68].

3.5
Mechanical Properties of Single Nanofibers

Single nanofibers produced by electrospinning can be expected to display quite unusual mechanical properties. The intrinsic structure of nanofibers as controlled by the freezing-in, crystallization, and elongational processes was already discussed above. Strong orientations of both chain molecules and crystals are known to affect mechanical properties such as the tensile stiffness and tensile strength considerably [69]. One reason is that these mechanical properties become more and more controlled by deformations of the covalent chemical bonds – i.e., their length and the angle between the bonds along the chain backbone – rather than by the corresponding deformations of van der Waals bonds acting between the chain molecules. A second reason is that the strong deformations of the polymers during electrospinning may give rise to modified crystal modifications, perhaps showing enhanced mechanical properties. Finally, it is known from other materials that a reduction in fiber diameter tends to enhance the strength of fibers as demonstrated by the so-called Griffin criterium [70].

Experimental investigation on the stress strain behavior of single electrospun nanofibers can only scarcely be found in literature so far, the analysis being mostly carried out via AFM methods. These studies very explicitly show that electrospun nanofibers possess very good mechanical properties. For example, moduli of up to 50 GPa and above were reported for polyacrylonitrile fibers [71]. The fibers showed high orientations according to X-ray analysis. Bulk samples of PAN without orientational order display moduli of only 1.2 GPa; the increase of the modulus by electrospinning is therefore quite significant. For polyethylene oxide nanofibers moduli were observed that were distinctly higher than the ones of bulk samples [72]. In this case again the orientation induced by electrospinning is the principal reason. In contrast to these results moduli of only 0.9 GPa were observed for nanofibers based on polyvinylpyrrolidone and containing titanium dioxide nanoparticles [73]. However, there are no statements on the orientational state in this case. It is in any case obvious that nanofibers feature very high moduli if the crystallite and chain orientation is high. Such fibers are thus of great interest among others for the application in nanofiber reinforcement.

4
Nonwovens Composed of Electrospun Nanofibers

Standard electrospinning devices such as the single or multiple syringe type die arrangements arranged opposite to a flat extended counter electrode (see Fig. 2 as an example) will give rise to so-called nonwovens displayed in Fig. 5. As far as the deposition of infinitely long fibers is concerned, fibers or segments of the fibers are deposited continuously within a certain area of the counter electrode or substrates on top of it. The layer thickness is controlled via the thickness of the fibers and the total coverage as specified, for instance, by the total weight of fibers per unit area. The planar extension of the nonwoven is controlled on the one hand by the bending instability giving rise to extended loops (see above) for a given electric field and on the other hand by the distance between the two electrodes. Taking a distance of 15 cm as an example the diameter of the deposited nonwovens tends to amount to 10–25 cm for a single syringe type die.

Such nonwovens also arise from electrospinning setups in which the feeding electrode consists of a wire or cylinders with spikes attached to them, again using a nonstructured counter electrode. A specific feature of such nonwovens that is already obvious from Fig. 5 and that will be discussed below in more detail is that it is highly porous. In fact, the total porosity, i.e., the volume taken by the pores relative to the one of the total nonwoven, may amount to 90–95%. The pore sizes are subjected to a significant distribution to be discussed below and the average pore dimensions tend to be different in the plane of deposition and perpendicular to it. Such nonwovens are of interest for filter and textile applications as well as for tissue engineering the filter efficiency, thermal insulation, or the growth of stem cells depending strongly on details of the pore structures. Furthermore, thin layers of such nonwoven may serve as surface coating, e.g., for modification of the wetting properties of the substrate all the way towards self-cleaning in analogy to the lotus effect.

4.1 Fiber Arrangement in Nonwovens

Electrospinning performed with planar counter electrodes, substrates located on them, and also moving substrates such as filter paper or textiles will give rise to a random planar deposition of the nanofibers (Fig. 5). Nonwovens thus result with an isotropic fiber orientation within the plane of the nonwovens and a layer-like arrangement of such nanofibers in the third dimension specifying the thickness. Yet, electrospinning is really not limited to the production of such nonwovens with a statistical planar orientation. The orientation of nanofibers along a certain direction is of interest for nanofiber reinforcement or for tissue engineering in order to allow a distinct growth direction for the cells and for many more applications to be discussed later.

Parallel fibers can, for example, be obtained using rapidly rotating cylindrical collectors which either serve as counter electrode or are combined with further electrode arrangements [64]. The collectors usually have the shape of long rotational cylinders but can also be wheel-shaped. Furthermore, parallel fibers can be realized by special electrode arrangements consisting of two flat plates aligned parallel to each other or corresponding frame-shaped electrodes, respectively [62, 74]. The degree of orientation of the fibers can be increased quite significantly by this method (Fig. 15). Another possible modification of the electrodes is a quadratic arrangement of four electrodes that leads to a cross-shaped deposition of nanofibers [75]. Further variations along this directions suggest themselves.

High degrees of orientations were achieved also by a distinct decrease of the distance between the electrodes down into the mm or even μm range in combination, for instance, with a finely tip-shaped spinning electrode or

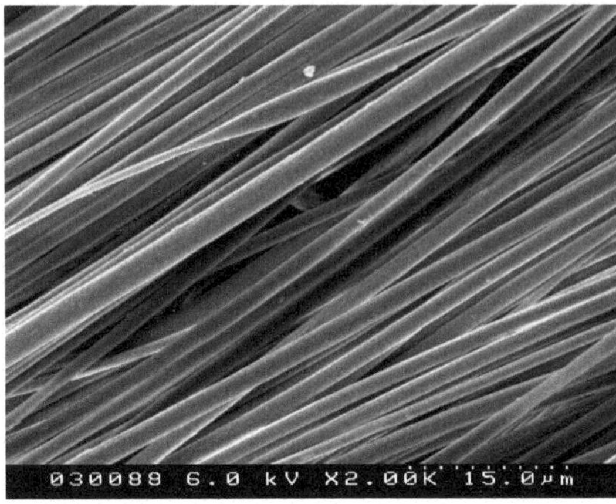

Fig. 15 Parallel PA 6,6 fibers obtained via electrospinning on rotating type electrodes

counter electrode respectively [76–78]. To obtain three-dimensionally oriented fiber arrangements within the nonwovens the needle punching technique known from conventional nonwovens made of macroscopic fibers or the solidifications using water jet treatment are possible means. Yet, there is no reference in literature referring to this technique as far as nonwovens made of nanofibers are concerned. Finally, it has to be pointed out that strong interest currently exists towards the formation of textile fibers via electrospinning.

4.2 Heterogeneous Nonwovens

It is generally the aim to prepare nonwovens with fibers of uniform thickness and from material of a particular chemical composition throughout the nonwoven. The reason is that in such case both the pore structure and the nature of the surfaces are uniform which allows better predictions for pore structure and adsorption properties (see below) and which may be of benefit for specific applications

Yet, it is also beneficial to a variety of applications to construct nonwovens with fibers of non-uniform thickness to, for example, introduce pore size gradients or modify transport properties. Experimental studies already discussed above and also theoretical treatments show that by proper choices of solvent and concentration of the spinning solution the fiber diameter for one and the same material can be varied by a factor well above 10. Good examples are polyamide or polyacrylonitrile, respectively. A possible approach towards nonwovens composed of nanofibers with different yet specified diameters consists in using a multi-jet electrospinning setup as represented by a parallel arrangement of several syringe type dies close to each other. Spinning solutions containing different concentrations of the same polymer, which are

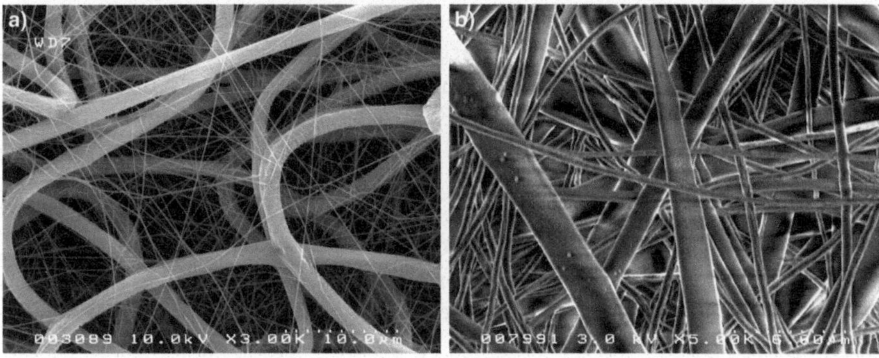

Fig. 16 Nonwoven composed of: **a** Thick nanofibers and thin PA 6 nanofiber, **b** Thick polyamide 6 (PA 6), and thin polylactide fibers [38]

fed through these dies, will give rise to nonwovens with nonuniform fiber diameters [38]. Two examples are given in Fig. 16, showing that the ratio of the fiber diameters within one nonwoven can be well above a factor of 10. Furthermore, by selecting the composition of the spinning solutions appropriately, one is even able to construct the nonwoven from fibers with spherical and flat cross-sectional shapes. This way, one becomes able to modify pore structures and adsorption properties correspondingly.

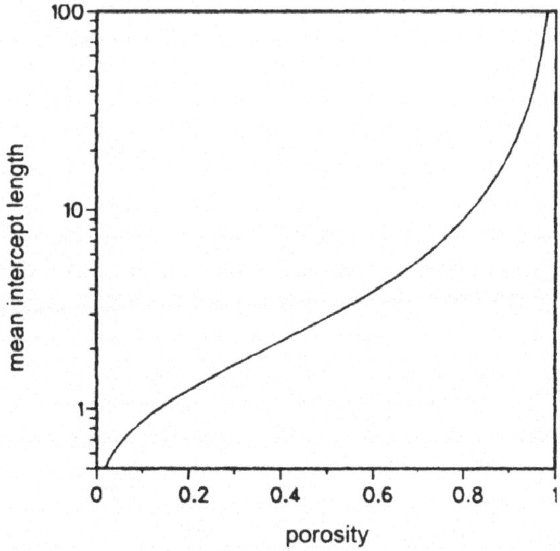

Fig. 17 Relation between reduced pore size (pore size relative to the fiber diameter) mean intercept length), and total porosity for arrangements of unimodal nanofibers, adapted from [81–83] as obtained from Monte Carlo simulations

Along the same line chemically inhomogeneous nonwovens can be produced with multi-jet arrangements accordingly by pumping polymer solutions composed of chemically different polymers through the dies, an example for a nonwoven is displayed in Fig. 16b. It is composed in this case of polyamide and polylactide nanofibers. In addition one can, of course, also control the fiber diameters of the two polymers with respect to each other, adding a further degree of inhomogeneity to the nonwoven.

Such chemically inhomogeneous nonwovens are of interest for specific applications. For tissue engineering, for example, heterogeneous carrier matrices based upon micro- and nanofibers, e.g., from PEO, collagen and segmented polyurethane, were fabricated by either sequential or simultaneous electrospinning [79]. A cylindrically shaped structured tissue based on thin collagen nanofibers as inner layer and thick polyurethane fibers as outer layer was found to constitute a good carrier matrix for artificial blood vessels.

4.3
Porosity and Pore Structures

The pore structures characteristic for these kind of fiber-based nonwovens determine the gas diffusion through the fibers, the resistance towards airflow, filter effectiveness, and the suitability as carrier for tissue engineering, e.g., for stem cells. So far only a limited number of experimental results on pore dimensions and internal surface areas were obtained. Scanning electron microscopical observations such as the ones displayed in Figs. 5 and 16 indicate that the pores are not spherical but rather anisometric assuming odd shapes as given by the multitude of linear fiber segments crossing each other, thus defining the surface of the individual pore. The average pore diameters are broadly distributed and are for fiber diameters of some 100 nm in a range between some 10^3 and 10^4 nm. However, these values can be distinctly modified by the control of the overall porosity defined as the volume taken by the pores relative to the total volume of the nonwoven. This parameter can, for instance, be determined from the total volume of a piece of nonwoven and its mass taking the density of the nanofiber material into account. More detailed information becomes available by using various types of porosimeters, e.g., mercury-porosimetry or nitrogen porosimetry detailing the total pore volume, average pore sizes, as well as the magnitude of the total specific surface area [80]. Such studies are currently extremely limited for nanofiber nonwovens.

In fact, the analysis of details of the pore structures of nonwovens particularly composed of nanofibers is not an easy task particularly if one is interested in their dependence on the orientational distribution of the fibers within the nonwovens. Planar isotropic arrangements, arrangements of parallel fibers in a given plane or fiber arrangements characterized by a three-dimensionally isotropic arrangement should result in different pore structures and corresponding permeation properties. It is of interest to learn how pore sizes are controlled by fiber diameters and perhaps at given diameters by the specific mechanical properties of the fibers, whether they be stiff or flexible.

Theoretical studies both analytical ones and simulations often dating back several decades have already proven themselves to be extremely helpful in this matter. In fact, such studies were motivated in the context of classical non-wovens composed of fibers with diameters well in the 10 micrometer range aiming at applications in filters or textiles. The predictions of such theoretical studies were displayed in reduced quantities, scaling with the fiber diameters so that the predictions can also be, to a certain extent, directly adapted to nonwoven prepared from nanofibers. One has to keep in mind, however, that the flow pattern may change significantly as the fiber diameter approaches the few nanometer scale as controlled by corresponding Knudsen numbers. The Monte Carlo approach taken for predicting nonwoven properties is particularly transparent since it mimics the deposition process taking place in electrospinning closely [81–83]. It will be used therefore in the following for the discussion of the pore structure of nonwovens.

In these Monte Carlo simulations linear fiber segments with a given fiber diameter are deposited on a model plane one after the other. Their individual orientations are chosen by various methods in such a way that either planar isotropic, planar parallel, or even three-dimensionally isotropic fiber orientations are mimicked. Furthermore, the fiber coverage per unit area can be specified, thus yielding nonwovens of different total porosity. The next step concerns the analysis of the pore structures. To this end quantities such as the mean intercept length can be calculated from the simulated structures. An intercept length typically represents the length of a linear segment oriented in a given way, which just fits into a pore bordered by the nanofibers. Within the framework of this paper the intercept length is given relative to the fiber diameter. The average is then taken with respect to all segment orientations and all positions in the nonwoven. Such a quantity can be taken as a good representation of the average linear dimension of a pore. Figure 17 displays a general correlation between the reduced fiber diameter and the reduced intercept length as a function of the total porosity.

One obvious result is that the higher the porosity the bigger the average pore size in nonwovens for fibers of constant radius. Values of the specific surface of nanofibers are typically in the range of 500 m^2/g and 10 m^2/g for nanofibers in the range of 10 to 500 nm. One rather surprising prediction is that as far as diffusion and permeation properties are concerned the orientational order of the fibers in nonwovens does not really play a significant role for larger total porosities as characteristic for electrospun nonwovens. Strong differences occur at small total porosities in agreement with the predictions of corresponding analytical treatments.

Monte Carlo simulations show furthermore that the overall porosities of nonwovens are significantly decreased with increasing flexibility of the fibers [84] (Fig. 6a) and that a chemical vapor deposition (CVD) [85] which leads to the construction of a shell around the fibers can distinctly decrease the porosity as well [86]. The simulations have been extended to predict the diffusional properties of gases in such nonwovens, the pressure drop in air filtration, the filter efficiency in aerosol filtrations, the permeation of viscous fluids, etc. [81–83, 87]

4.4
Mechanical Properties of Nonwovens

The mechanical properties of the nonwoven are of interest if, for example, they are used as template for seeding stem cells in tissue engineering. The mechanical properties should be similar in this case to the ones of the tissue to be replaced. Tissues such as cartilage tissue for example or also skin tissue are typically characterized by elongation moduli (stiffness) ranging between some 10 to some 100 MPa, by maximum deformation stresses (strength) of some 10 MPa and maximum deformations of about 10 to 200%. To obtain corresponding data for nonwovens made of nanofibers these nonwovens are subjected to stress-strain investigations, for instance, with commercially available mechanical testers. For electrospun nanofibers which do not show a specific molecular orientation along the fiber axis and which are based on soft elastomers values for the modulus of about 3 MPa, for the maximum stress of about 9.6 MPa and for the maximum elongation of about 360% are reported [88]. In contrast nonwovens made of significantly stiffer polyamide fibers (Fig. 18) display

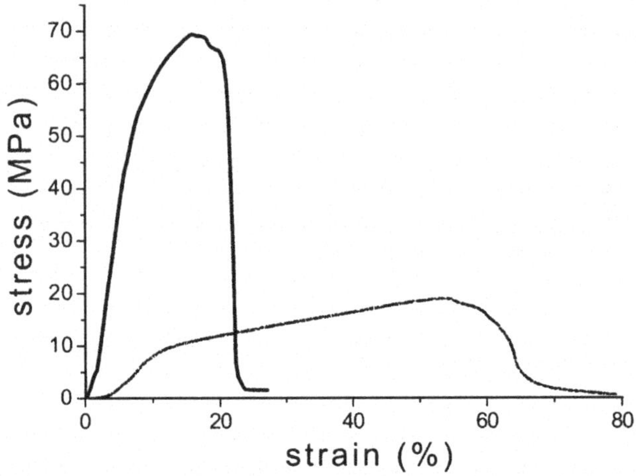

Fig. 18 Stress-strain experiments performed on polyamide 6/6T nanofiber nonwovens [89] with the fibers either parallel (*blue, upper curve*) or perpendicular (*red, lower curve*) to the deformation direction

values for the modulus of about 100 MPa, the tensile strength amounting to 20 MPa and the maximum elongation to 53% [89]. It is furthermore evident that nonwovens in which the fibers are oriented along a preferred direction within a plane should exhibit enhanced mechanical properties. In fact, the modulus increases due to fiber orientation from 100 to 900 MPa for PA 6 fibers oriented along the elongation direction. The tensile strength is increased from 20 to 70 MPa and the elongation at break is reduced from 53% to 18% [89]. For electrospun fiber nonwovens acting as scaffold for tissue engineering and prepared from collagen with fiber diameters of about 100 nm moduli of 52 and 26 MPa for elongation along and perpendicular to the orientation direction, respectively, were measured for oriented fiber arrangements with the strength amounting to 1.5 and 0.7 MPa, respectively.

5 Recent Electrospinning Developments

Electrospinning, the resulting nanofibers and nonwovens composed of these nanofibers quite obviously offer unique opportunities in terms of functional nanoobjects and corresponding applications. Yet, despite the richness of features characteristic for electrospinning and the resulting structures the approaches discussed in detail so far may meet with limitations for specific areas of application. Extensions of the basic electrospinning techniques were developed for this reason characterized by other kinds of self-organizations. In particular, these are electrospinning with strongly reduced electrode distances as well as co-electrospinning. Furthermore, electrospun fibers can be used as template to produce hollow fibers.

5.1 Electrospinning with Strongly Reduced Electrode Distances

The patterning of surfaces via a spatially controlled deposition of nanofibers as well as a deposition of nonwovens with well defined geometry and orientation meets with increasing demand. Possible applications include nanofluidics, superhydrophobic patterning, and the integration of nanofibers or nanofiber assemblies into hierarchical structures comprising nanoelectronic or nanophotonic circuits. Among others, current approaches include the design of appropriate counter electrode configurations or the manipulation of the nonwovens composed of such nanofibers once it has been prepared by electrospinning. What is needed is an approach which allows a direct and precise deposition of nanofibers or nanofiber assemblies onto the designated target area.

More recently the so called near field electrospinning was proposed [76–78]. A droplet suspended at an AFM tip is subjected to an electric field existing between the tip and a counter electrode which is located at a distance well below the micrometer range [76]. The droplet is deformed as expected and subsequently deposited in the shape of a rod/short fiber on a moving substrate. The scale is in the nanometer range as the droplet diameter is also in the nanometer range. A further approach involves a microfluidic feeding setup with very fine tips and a rapidly moving take-up substrate [77]. A larger scale technical realization devoted to continuously patterning of extended surfaces or to the continuous incorporation of nanofibers/nanorods into particular nanoelectronic devices is not obvious at this stage.

Another approach working in a continuous fashion – called high precision deposition electrospinning (HPDE) – is based on the experimental die-counter electrode setup used for conventional continuous electrospinning and consists in reducing the distance between the two electrodes from the usual 10–20 cm down to a range extending from 100 μm up to a few millimeters [78]. To induce a continuous spinning process under such conditions, one often has to monitor the spinning process by optical means and to couple the optical signals to the control of the electric field. A reduction in distance will certainly give rise to a decrease of size of the area in which the nonwoven is deposited. Yet, one might also expect that the reduction in electrode distances gives rise to a suppression of the tendency towards the onset of the bending instability and the corresponding strong reduction in fiber diameters. Furthermore, the deposition of wet fibers which subsequently merge to films or are decomposed via Rayleigh instabilities to droplets might lead to problems. The surprising finding was that one becomes able to deposit individual nanofibers or nonwovens in a highly controlled way both in terms of small fiber diameters and of the deposition location and orientation as shown in Fig. 19. Various types of fiber depositions (straight or with loops, with

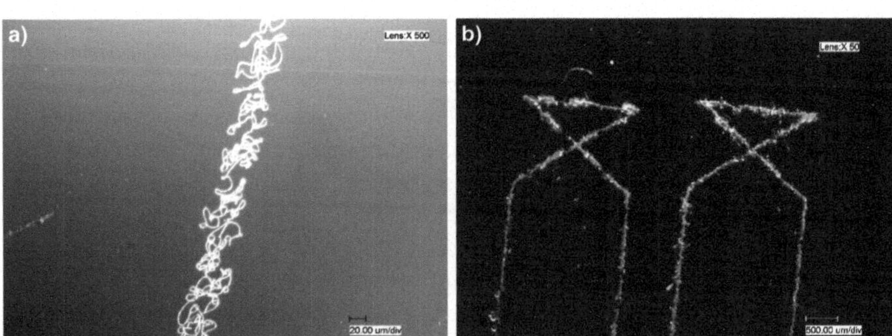

Fig. 19 Precision deposition of nonwoven pattern composed of PEO via the precision deposition approach [78]. **a** Linear trace of nonwoven, **b** Writing of a test pattern

droplets incorporated or without them) can be deposited in a controlled way, and one is able to "write" specific figures, or pattern by moving the substrates accordingly, as also shown in Fig. 19. The expectation is that this type of precision deposition will give rise to novel applications among others in areas relying on surface properties such as sensorics, microfluidics, and possibly the modification of surfaces of implants.

5.2 Co-Electrospinning

Co-electrospinning has been developed to produce core/shell fibers composed of polymer shells and cores, of low molar mass materials as core and a polymer shell, or even hollow fibers. Two dies arranged in a concentric configuration are connected to two reservoirs containing different spinning solutions in this approach. Compound droplets are formed at the tip of the two dies and deformed in the electric field. A compound jet is formed and a compound core-shell fiber is deposited at the counter electrode or substrates located on top of it [90]. The theoretical analysis has revealed that the entrainment of the core droplet may pose a problem. One possible solution is to choose the relative position of the interior die relative to the outer die appropriately [91].

Important features of co-electrospinning with respect to the incorporation of biological objects are: first, that the core can be a low molar mass fluid, including water or oil, providing a natural environment for the biological object, and, second, that the electric charges are located only at the outer surface, so that the inner droplet, and thus biological objects dispersed in it, do not carry charges at all. Yet, they are affected mechanically by the viscous stresses in the jet during spinning. Predictions of these forces are available from theory [91]. The estimate is that biological subjects such as proteins, viruses or bacteria might experience dangerous viscous stresses in specific cases and yet can be adjusted by suitable means to stay moderate.

Co-electrospinning has been applied so far for the preparation of polymer core shell fibers, hollow polymer core shell fibers, hollow fibers composed not only of polymers, but also of ceramics, as well as for the immobilization of functional objects in droplets dispersed in the core that are arranged along the fiber axis. Among the examples reported in the literature are: core shell fibers spun from polystyrene and polyethylene oxide, two kinds of polyethylene oxide (one with and the other without a chromophore), and core shell fibers with the electrically conductive polymer polyhexathiophene and the insulation polymer polyethylene oxide (Fig. 20). Hollow core shell fibers in which one polymer (polycaprolactone) forms the shell onto which the core material is deposited (polyethylene oxide) as inner wall is another example for the broad spectrum of fiber architectures which can be produced by co-electrospinning. The formation of the two-layer hollow fiber is based on the

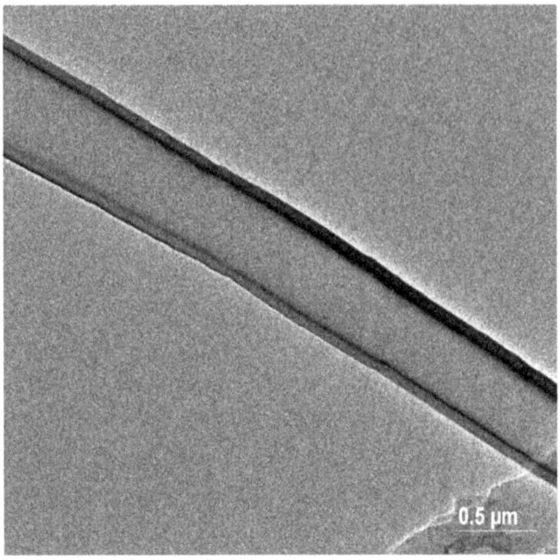

Fig. 20 Core shell nanofibers produced by co-electrospinning PVDF core/PC shell [90]

evaporation of the core solution through the shell yielding the deposition of the core material onto the inner shell layer.

Fluorescent dyes and biological objects such as the green fluorescent protein (GFP) were dispersed in core shell fibers in which the core region consists of water and the functional units. The architecture may be characterized in this case by droplets that are regularly arranged along the fiber axis as pearls in a necklace [92, 93]. Such fibers can be used, for instance, for sensorics: the GFP reversibly looses its capability to display fluorescence if in contact with denaturizing compounds such as urea. Furthermore titanium hollow fibers were produced using in this case a precursor route. The core materials were chosen to be a mineral oil, and the shell material consisted of a mixture of polyvinylpyrrolidone and the precursor compound TiO [94]. The removal of the oil from the core shell fibers followed by a calcinations step yields titanium hollow fibers.

The interest in co-electrospinning is continuously increasing, while applications related to tissue engineering, microfluidics, optoelectronics, and sensorics are being discussed in the literature. It should finally be pointed out that a coaxial jet generation has already been considered previously in the context of electrospraying of compound droplets [95].

5.3
Core Shell Fibers and Hollow Fibers via Templates (TUFT Approach)

Another concept on how to produce core shell fibers or hollow fibers is based on the exploitation of electrospun nanofiber as a template (TUFT: tubes by fiber templates) onto which a second material is deposited as shell either from solution or melt, via layer by layer techniques or from the vapor phase [96–99]. It is frequently preferable to apply the shell from the vapor phase rather than from a fluid phase since the latter may give rise to mechanical forces causing a disruption of the tiny template fibers towards an inhomogeneous coating or to a swelling of the template disturbing the original dimension of it.

Metals, glasses, or ceramics have been deposited as shell material from the vapor phase onto the template fiber and polymers, though the vapor phase deposition is limited in this case to very special polymers. Examples are polyparaxylylene (PPX) and its derivatives. A precursor, paracyclophane, and corresponding derivatives are pyrolized at 650 °C in this case yielding activated monomers which polymerize on deposition at room temperature on a substrate. One particular feature of the TUFT approach is that several layers of different materials can be deposited yielding nanofibers of complex architecture. The core shell fibers can be functionalized by incorporating nanoparticles and functional molecules such as chromophores, drugs, or catalysts into the core fiber via electrospinning. The shell layer is then being used either to control the kinetics of delivery of the functional molecules or to immobilize them, at the same time allowing them access to the functional molecules from the environment (see below). PPX has particular advantages as shell material for such applications, since it is insoluble in most solvents even at higher temperatures; yet, specific molecules are able to diffuse through PPX-layers via the amorphous regions.

To obtain hollow fibers the core fiber is removed by thermal decomposition, selective solvents or by biodegradation. The internal surface of the hollow fibers is structured in this case by the surface topology of the template fibers. Porous template fibers, for instance, give rise to hollow fibers showing an internal assembly of spikes. Among other reasons, this is of interest for microfluidic applications, since these spikes will affect both the wetting

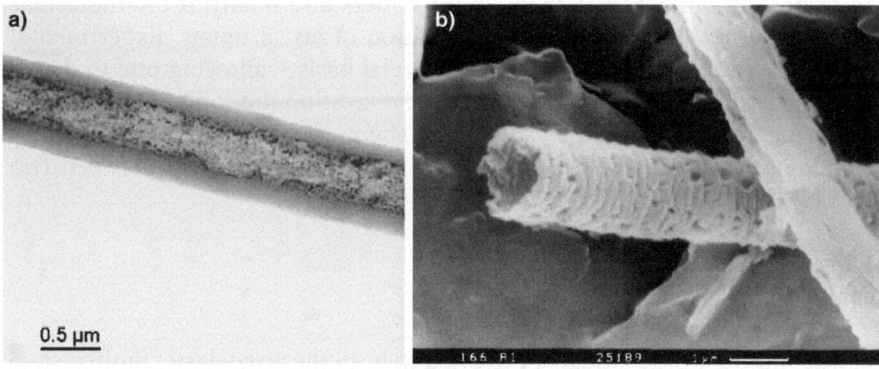

Fig. 21 **a** Hollow fibers composed of PPX and palladium nanoparticles produced via the TUFT-approach [99], **b** Aluminium oxide hollow fibers

of the nanotubes by a given fluid and its transport through the nanotube. Using the TUFT approach hollow fibers made from polymers and metals such as aluminum, chromium, copper, gold, nickel, and others composed of titanium dioxide have been prepared (Fig. 21). Titanium tetraisopropylate was deposited in another example as precursor material onto a polyamide template fiber, the template fibers were pyrolized, and the precursor was calcinated yielding titanium dioxide [98]. It is obvious that the TUFT-approach broadly extends the range of applications of electrospun fibers.

6 Brief Review of Materials Which have been Electrospun to Fibers

6.1 Spinning of Technical Polymers

Electrospinning is currently applied predominantly to polymers of synthetic and natural origin. The major reason is that polymer solutions and polymer melts possess specific flow properties that support the fiber forming process in electrospinning taking place via self-assembly. These flow properties have their origin in the particular conformation of long chain molecules – statistical Gaussian coils have sizes of the order of several 10 nm (as far as long flexible chain molecules are concerned) – and their interaction with each other via entanglements in melts or more concentrated solutions [100, 101]. Such chain entangled fluids are characterized by very unique viscoelastic properties [37], and they are, in turn, able to sustain the strong elongational strains, strain rates and the looping and spiraling pathways characteristic of the bending instability in electrospinning without rupture. A decrease of the molecular weight of a given polymer decreasing towards oligomers, keeping the spinning parameters constant, tends to give rise to the formation of beaded fibers rather than smooth fibers and finally, if the molecular weight is further decreased, to the formation of just droplets. Experimental methods are available – also on a commercial basis – allowing one to determine the viscoelastic response, viscoelastic relaxation times of polymer melts, and, in particular, of polymer solutions. The analysis is based on the time-dependent changes in the diameter of a fluid meniscus located between two plates [37].

6.1.1 Spinning from Polymer Melts

Electrospinning can obviously be applied to both the viscoelastic molten state as well as to the viscoelastic solution state. Yet, so far only a very small number of the studies have been published in the literature considering elec-

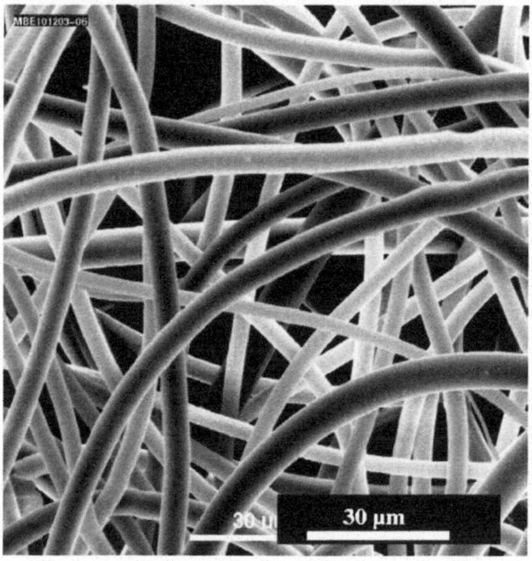

Fig. 22 PA 6 nanofibers produced by melt electrospinning [106]

trospinning from the molten state [102–106]. This holds despite the fact that two topics of great technical importance, i.e., the need to get rid of the large amount of solvents involved in solution electrospinning and the need to increase the production rate, are met by melt electrospinning. The problem which melt spinning still encounters is that the resulting fibers tend to possess diameters well outside the nanometer range, i.e., of the order of a few micrometers up to several 10 µm. The high viscosity and the peculiar viscoelastic properties of melts as controlled by the exceedingly dense entanglement network formed by the polymer chains are two reasons for this problem. Electrospinning has, in any case, to be performed at elevated temperatures well above the respective melting or glass transition temperatures. Strong electric fields are beneficial in this context and, in fact, electrospinning has been done even in vacuum or in the presence of electrically protective gases such as SF_6 to achieve an electrical field as high as possible without encountering an electric breakthrough.

Polypropylene, polyamide 12, polyethylene terephthalate, polyurethane, polycaprolactone polyamide 12 are examples of polymers which have been electrospun from the melt. In general the fibers' diameters are still in the micrometer range with PA 6 fiber diameters known today to be as low as 900 nm. Fibers with diameters even down to about 300 nm and below were more recently reported for the electrospinning of a low melting polycaprolactone [105]. A reduction of molecular weights seems a promising approach although properties of the solid materials tend to become poorer in this case. Nevertheless these results indicate that one may be able to produce nanofibers from polymer melts by tailoring polymer melt properties appropriately, for instance, via a careful selection of the molecular weight and its distribution as well as of the processing temperature relative to the glass transition or melting temperature. It is obvious that melt electrospinning is still in its infancy. The progress which has been made recently nevertheless indicates that one will succeed in the near future to produce electrospun fibers from the melt with diameters down to a few 100 nm and even below. This would give a major push to technical applications of electrospun nanofibers.

6.1.2 Electrospinning from Organic Solvents

Electrospinning from polymer solutions can be much more easily controlled, since the viscoelastic properties, surface tension, or even electric conductivity can be modified via the choice of the solvent, solvent mixture, the polymer concentration in the spinning solution, the temperature of the spinning solution, and, in particular cases, also via specific additives. It has been demonstrated that polymer fibers with diameters created to a few nanometer can be achieved by a suitable choice of the spinning parameters. An extremely broad range of technical, functional synthetic and natural polymers has thus been spun to nanofibers and the number of novel polymer systems produced as nanofibers is ever increasing [9–13].

Among the large number of polymers that have been electrospun from organic solvents to nanofibers are polystyrene (PS), polyacrylonitrile (PAN), polycarbonate (PC), aliphatic and aromatic polyamides (PA), polyimides (PI), polybenzimidazol (PBI), poly(ethylenterephthalate) (PET), poly(trimethylenterephthalate) (PTT), poly(hexamethylenterephthalate-*co*-hexamethylen-2,6-naphthalate) (P(HT-co-HN)), polyurethanes (PU), poly(ethylene-*co*-vinylacetate) (PEV), polyvinylchloride (PVC), polymethylmethacrylate (PMMA), polyvinylbutyral (PVB), cellulose acetate (CA), polyvinylidenfluoride (PVDF), and many more as detailed in several review papers [9–13]. Poly-(*p*-phenylenterephthalate) (PPTA) also known under the trade names Kevlar and Twaron) is an aromatic polyamide that cannot be melted without decomposition, yet it can be electrospun from appropriate solvents and at sufficient concentrations for which lyotropic solutions are formed, yielding fibers with extremely high tensile strengths. Other polymers with high thermal stability that cannot be melted without decomposition are PI and PBI. These polymers can be prepared as nanofibers via electrospinning of soluble precursors, which are subsequently transformed to the polymer species by polymer analogous reactions. Electrospinning of polyacrylonitrile fibers has been considered in detail. One reason for this consideration is that they can be converted into carbon fibers by pyrolysis. With electrospinning of poly(ferrocenyldimethylsilane), it was possible to show that organometallic polymers can also be spun into very fine fibers.

The advantage of electrospinning from organic solvents is the broad scope of polymer systems soluble in organic solvents. Disadvantages are, as pointed out above, the properties of the organic solvents themselves, such as flammability, toxicity, corrosiveness, disposal, etc. These disadvantages may not be an issue on a laboratory scale but will play a major role in industrial production setups. Spinning of fibers from water soluble polymers constitutes a major benefit in this respect.

6.1.3 Electrospinning of Water Soluble Polymers

Water is an ecologically preferable and safe solvent, its solubility can be adjusted by variations of the pH-value or the temperature, for instance. A major disadvantage is, of course, that fibers composed of water soluble polymers tend to decompose rapidly upon contact with water. While this may be positive for biomedical applications, additional stabilization of these fibers, among other techniques, by cross-linking, is necessary for other technical applications, such as filters or textiles.

Extensive investigations have been carried out on water soluble polymers such as polyethylene oxide PEO, polyvinylacetate PVA, poly(acrylic acid) (PAA), poly(acrylic amide), polyelectrolytes, poly(vinylpyrrolidon) (PVP), and hydroxypropylcellulose (HPC). PEO is especially versatile, because it is also soluble in many organic solvents. Therefore, it can be electrospun from a variety of solvents. Compared to PEO polyvinylalcohol offers an even larger scope of variations. Starting from polyvinylacetate the degree of hydrolization of PVA can be adjusted and thus the solubility of the polymer in water. The hydroxy groups present in PVA can be used either before or after electrospinning to create chemical reactions for multiple purposes. PAA can be electrospun from aqueous solution almost just as well as PVA. Further modifications are possible by variation of pH-values and addition of salts, respectively. The physical stability of PVA/PAA blends towards water can be significantly improved by chemical cross-linking reactions. The water resistance of electrospun PVA/PAA blend fibers can be increased by analogous esterification reactions or by aldol reactions with polyaldehydes. An interesting alternative to cross-linking of PVA with PAA or polyaldehydes can be achieved with electrospun PVA/cyclodextrin fibers that show a strong pH-dependent swelling behavior versus water after thermal cross-linking. Chemical cross-linking of electrospun PVA derivatives can also be accomplished for PVA functionalized with photo-crosslinkable substituents via exposure to light.

One alternative for the cross-linking of electrospun fibers is the so-called "reactive electrospinning". In an example a mixture of 2-hydroxymethylmethacrylate, methacrylic acid, ethylenglycoldimethacrylate, 2,2′-diazo-*iso*-butyronitril and a photo cross-linking agent were first pre-polymerized and subsequently photochemically cross-linked during the electrospinning process. Following a similar concept poly(dicyclopentadien) fibers can be ob-

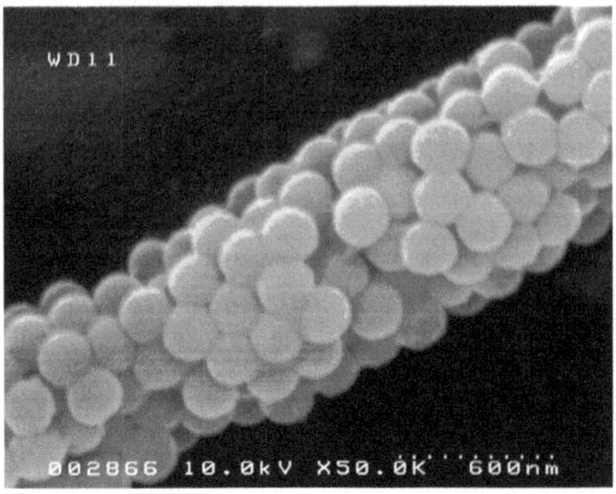

Fig. 23 Nanofibers based on polystyrene latex particles spun from water dispersions yet being water insoluble as electrospun fibers

tained by "reactive electrospinning". Finally, polyvinylpyrrolidone, which can also be electrospun from aqueous solution, was used to produce structured electrospun fibers by spinning of PVP blends and subsequent selective extraction.

A highly interesting route towards water stable nanofibers using water as solvent is based on colloidal dispersions of polystyrene latex particles [107]. To obtain fibers during electrospinning a water soluble polymer such as PVA has to be added to the spinning solution in low concentrations (typically about 6 wt. %), which is subsequently removed after fiber formation by rinsing the fibers in water. The latex particles fuse within the spun fibers yielding water insoluble nanofibers. Figure 23 shows the resulting fibers, which in fact are no longer soluble in water. It is very obvious that the fibers are composed of nanoscalar spheres. This approach can be extended to other types of polymer species.

6.2
Spinning of Biopolymers

Electrospun nanofibers, nonwovens composed of biopolymers are of ever growing interest in medicinal applications, including surface modifications of implants, tissue engineering, wound healing, and drug delivery. Some of these areas will be discussed below in more detail. A set of biopolymers, modified biopolymers, and their blends with synthetic polymers were processed into nanofibers by electrospinning often only under very specific conditions, such as using special solvents or processing these materials as blends with synthetic polymers such as PEO or PVA.

Collagen can be electrospun from hexafluoroisopropanol or as blend with PEO, PCL, and PLA-co-PCL leading to collagen fibers with diameters of 200 to 500 nm and denatured collagen, so-called gelatin, from aque-

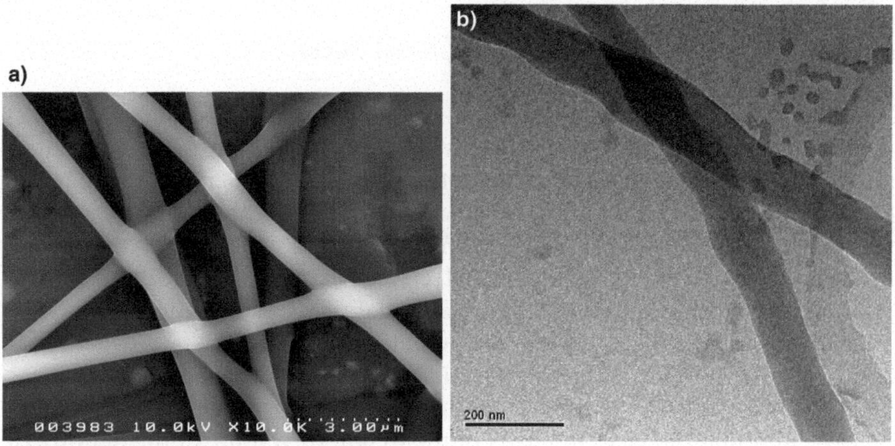

Fig. 24 Nanofibers from: **a** Collagen, **b** Chitosan

ous solution and trifluorethanol, respectively, as pure polymer or as blend (Fig. 24a) [108]. Chitin and chitosan can be electrospun to nanofibers in its pure form and as blends as well (Fig. 24b) [109]. These fibers are of particular interest for wound dressings. Other proteins and enzymes such as casein, lipase, cellulose, bovin serumalbumin, and luciferase can only be processed by electrospinning as blends with synthetic polymers. A series of papers have described electrospinning of silk and silk-like polymers mainly for biomedical applications [110]. Cellulose as classic fiber material can be electrospun from *N*-methylmorpholin-*N*-oxide/water systems, and dimethyl acetamide/LiCl systems, respectively, into fibers with diameters in the submicrometer range. Cellulose acetate as organosoluble cellulose derivative and well established filter material was electrospun to fibers without any problem.

Polymers which can be hydrolyzed under physiological conditions or are biodegradable (summarized as bioerodible polymers), such as aliphatic polyesters, polyanhydrides, polyphosphazenes, etc., are very important for a large variety of applications. Electrospun fibers of such bioerodible polymers are intensely studied for pharmaceutical applications or for the field of regenerative medicine (tissue engineering). Polylactide (PLA), an aliphatic polyester, is one of the classical bioerodible polymers that were successfully electrospun. One reason for the easy processing of the different PLA isomers by electrospinning is their high solubility in halogenated aliphatic solvents. Polyglycolid, in contrast, was used in electrospinning processes only to a minor degree. Poly(ε-caprolactone) (PCL), which is easily soluble in different solvents, and PCL copolymers were extensively used for fabrication of electrospun scaffolds. An interesting alternative to PLA and PCL are poly(hydroxybutyrate)s and their derivatives.

6.3
Nanofibers from Polymer Hybrids, Metals, Metal Oxides

Nanofibers/nanowires composed of nonpolymeric materials, such as metals, metal oxides meet with strong interest for applications in nanoelectronics, nanooptics or nanomagnetism due to the specific dependence of their properties on the 1-d character of fibers, on the diameter of the fibers, and on their axial ratio [111–113]. Various methods are now available to produce nanowires including among others electron beam methods, interferometric lithography, microprobe assisted manipulations, or electrodeposition into porous templates, such as aluminum templates prepared by electrochemical approaches or into blockcopolymers [112]. Disadvantages, however, of these methods include that they are in the majority of cases time consuming, expensive, and that the yield in terms of the weight or volume of the nanowires that can be produced is very limited. Electrospinning offers, in principle, a highly attractive alternative.

Yet, a major problem as far as the application of this technique for the production of metal or oxide fibers for that matter is that – as already pointed out above – solution or melt appropriate for electrospinning should display specific viscous, in particular viscoelastic properties. Such specific rheological properties are natural for polymer materials and it is for this reason that the majority of papers dealing with electrospinning have concentrated on polymer materials. Yet, electrospinning can be exploited for the preparation of metal, metal oxide, semiconductor nanowires and of the corresponding polymer hybrids following the precursor route. Precursors of the target materials such as metal salts are introduced for this purpose into spinning solutions containing polymers and an appropriate solvent able to solve both the polymer and the salts. Electrospinning of such a ternary solution is found to give rise to polymer nanowires in which the salts are homogeneously dispersed provided that the spinning parameters are chosen appropriately.

The precursor materials are, in a subsequent step, reduced to the target metals or metal oxides via thermal treatment or the treatment in the presence of reducing agents such as hydrogen. The results are polymer hybrid nanofibers. In a further step the polymer can be removed completely, in general again by thermal treatment, for example, to yield nanofibers from the pure metal or metal oxide material. All of these steps can be done on a technical scale yielding sizable amounts of nanowires.

6.3.1
Polymer Hybrids

Polymer hybrid nanofibers in which nanoparticles, composed of platinum, rhodium, and palladium, are dispersed were prepared along the precursor

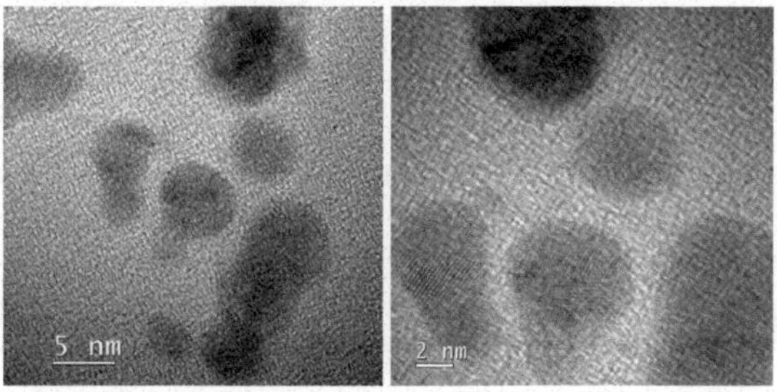

Fig. 25 Rhodium-palladium bimetallic nanoparticles for catalytic applications grown within the nanofibers via the precursor route [114, 115]

route using appropriate salts, aiming at heterogeneous catalysis [114, 115]. To obtain bimetallic particles the spinning solution has to be composed of the two corresponding precursor salts in addition to the polymer species and the solvent. The reduction yields nanoparticles within the polymer nanofiber with diameters typically in the 3–5 nm range, which are again highly dispersed (Fig. 25). An EDX analysis revealed that all bimetallic particles observed by TEM are composed of the two metals in the ratio defined by the concentration of the precursor salts in the original spinning solution. An interesting feature of the approach described here is that bimetallic nanoparticles become available with compositions that are not accessible on a macroscopic scale for which the observed compositions are located well within the binodal regime. Catalytic test reactions with such polymer nanofibers were highly successful.

In a similar way polymer hybrid nanofibers were produced containing nanoparticles with either ferromagnetic or superparamagnetic properties [115–117]. Fe_3O_4, for example, is a well known ferrimagnetic material

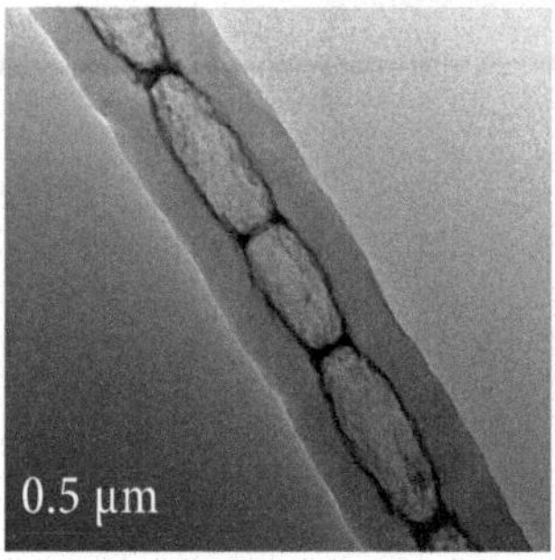

Fig. 26 Compartimented nanofibers composed of PPX with superparamagnetic iron oxide particles at interfaces [115]

which, however, becomes superparamagnetic as the size of the particles is reduced down to the nm-scale. Superparamagnetism is characterized by a very high magnetization in the presence of an external magnetic field, yet the magnetization is not permanent and no hysteresis effects characteristic of ferromagnetism are observed. Highly interesting morphologies are developed if the original nanofiber is coated by PPX using the TUFT-approach in order to stabilize the nanofibers mechanically and thermally, and if the reduction is subsequently performed at an elevated temperature [115]. The polylactide melts in the selected temperature range; the polylactide core fiber then tends to fragment into nanodroplets via the well known Rayleigh instabilities. The final result is a compartmented fiber as shown in Fig. 26 in which the nanoparticles have accumulated on the surface of these droplets. This enrichment of nanoparticles at interfaces – colloidal jamming – has been the subject of several papers in the literature. The fibers nevertheless display superparamagnetism despite the agglomeration at interfaces due to their separation by polymer sheets.

6.3.2 Metal and Metal Oxide Nanofibers

Metal/metal oxide nanofibers become available if one succeeds in dispersing high amounts of the precursor species in the spinning solutions well above a weigh ratio precursor/polymer of 50% up to 80% and more [115, 118, 119]. The preparation of nanofibers composed of copper is an example. Copper nitrate was chosen as precursor salt. It was dispersed in a spinning solution containing the solvent mixture water/isopropanol and polyvinyl butyral as polymer component. Electrospinning gave rise to homogeneous nanofibers with diameters in the 550 nm-range. The thermal reduction and a subse-

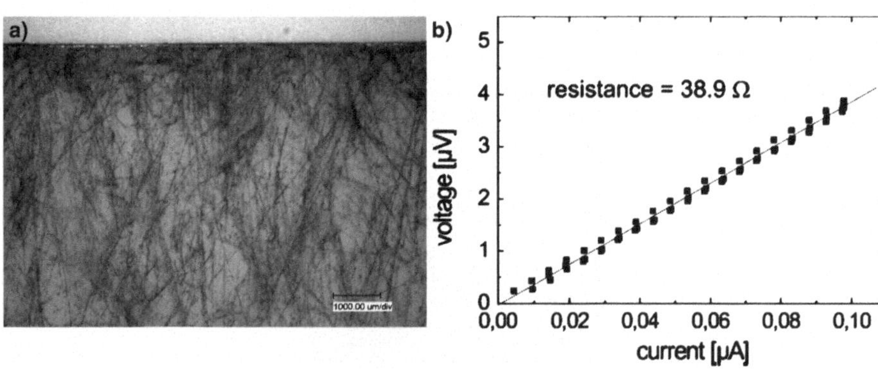

Fig. 27 **a** Nanofibers composed of Cu prepared via the precursor route and **b** single fiber current/voltage characteristics [118]

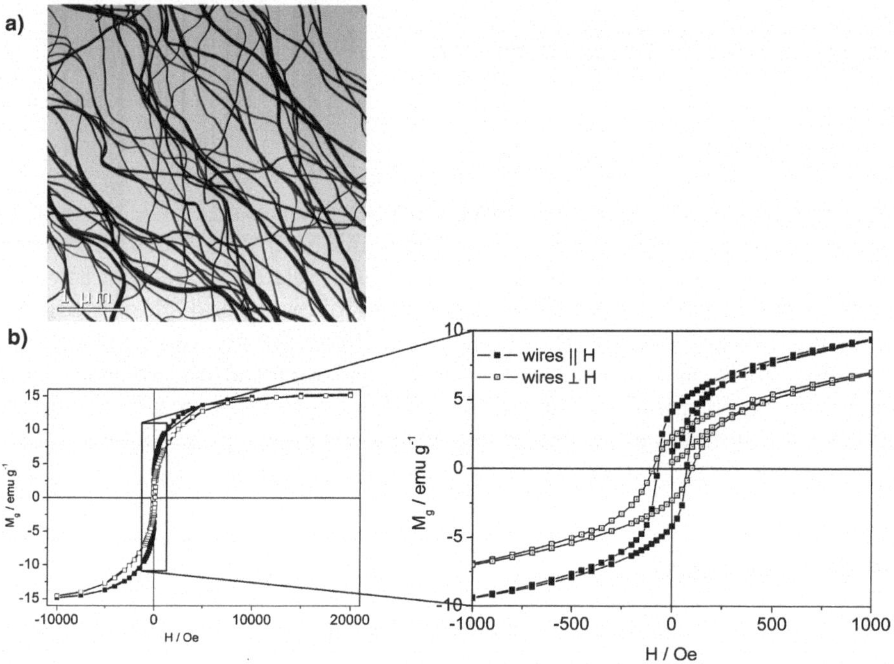

Fig. 28 **a** Co nanofibers and **b** Anisotropic magnetization diagram of ferromagnetic Co nanofiber [115, 119]

quent reduction in the presence of hydrogen yielded nanofibers composed of pure copper (Fig. 27) displaying the reddish color characteristic of copper and the X-ray analysis showed the crystal structure characteristic of copper. The diameter of the copper nanofibers amounted to about 270 nm, which is well below the diameter of the original fiber primarily due to the loss of the polymer component. Electrical conduction investigations revealed a specific conductance of $8.5 \times 10^3\ \Omega^{-1}\ \mathrm{cm}^{-1}$.

In a similar way nanofibers composed of nickel, iron, iron oxides, or cobalt were prepared, displaying specific magnetic properties [115, 119]. Cobalt nanowires are shown in Fig. 28, as an example, the radius being controlled from below 20 nm up to several 100 nm depending on the salt loading of the spinning solution. Nanofibers composed of ferromagnetic materials are expected to show ferromagnetism rather than superparamagnetism even for very small fiber diameters, since they are elongated in one dimension well above the nanometer-scale [112, 113] and in fact they display anisotropic magnetization diagrams as shown in Fig. 28. The hysteresis displays different shapes for the magnetic field oriented along the fiber axis as compared to the perpendicular case.

7 Applications for Electrospun Nanofibers

7.1 Technical Applications

The spectrum of applications that can be envisioned for electrospun nanofibers is extremely broad due to their unique intrinsic structure, surface properties, and functions. Applications both related to Material Science and Life Science areas will be discussed in the following, though only briefly since the focus of this contribution is primarily on the processes controlling nanofiber formation in electrospinning by self-organization processes. For each of the applications mentioned in the following the specific set of features of electrospun fibers of particular benefit for this application will be pointed out.

7.1.1 Template Applications

Electrospun polymer fibers can be used as examples for the preparation of hollow fibers (tubes by fiber templates process = TUFT process) (e.g., see [97–100]). According to the TUFT process electrospun biodegradable or soluble polymer fibers are coated with polymers, metals or other materials. After selective extraction or degradation of the template fibers hollow fibers whose dimensions depict the negative replication of the template fibers are obtained. Au, Cu and Ni hollow fibers were fabricated in an analogous way. In a further development of the TUFT process hollow fibers with intricate architectures were prepared by electrospinning using the layer by layer technique. Both, the magnitude of the inner diameter as well as the surface topology of the interior wall are controlled by the template fiber properties, and the template fiber allows the introduction of functional materials into the hollow fibers.

7.1.2 Textile Applications

Nanofiber based nonwovens can be used to strongly modify the properties of conventional textiles composed of much thicker textile fibers (e.g., see [121,122]). The nanofibers allow a great increase in wind resistance, ability to adjust water vapor permeability, optimize thermal isolation behavior, and to prepare textiles with special functionalities (such as the self-cleaning (lotus) effect, aerosol filtering, and protection against chemical or biological hazards). Features which control these functions are predominantly the intrinsic pore structures, the corresponding permeation properties for gases and fluids, and the high specific surface area that comes with the nanodimensions.

7.1.3
Filter Applications

High filter efficiencies rely on the fact that the channels and pores within the filter material can be adjusted to the fineness of the particles which are supposed to be filtered out. By the transition from fibers with diameters in the μm range to fibers with diameters in the nm range one becomes able to filter finer and finer particles (e.g., see [122,124]). Coalesence and aerosol filters are characteristic applications for nanofiber nonwovens. Very recently aerosol filtration was simulated for nanofiber filters concentrating on reduced operating pressures [123]. It was pointed out in this paper that the flow lines around nanofibers may differ strongly from the ones around μm-sized fibers, in which case the free molecular flow regime may become dominant. Important results of the simulations are that the collection efficiency of the nanofiber filters increase strongly with decreasing fiber diameter at constant pressure drop and that the diameter of the particle which can be captured with high efficiency also decreases strongly with decreasing fiber diameter. Furthermore, for gas filters a transition from depth filtration to surface filtration can be induced which allows the filter to be cleaned by reverse pressure pulses very effectively increasing the lifetime of such a filter by up to a factor of 10.

7.1.4
Catalysis

A huge problem in catalysis is the removal and recycling of the catalytic agent after the reaction (e.g., see [115,116,125]). The implantation of homogeneous as well as heterogeneous catalysts into nanofibers poses an interesting solution for these problems. The reaction mixture can circulate around the fibers as is the case in the continuously working microreaction technique, or the fibers fixed on a carrier can be immersed repeatedly into a reaction vessel to catalyze the content of the vessel. Short diffusion distances within nanofibers and the specific pore structures and high surface areas of nanofiber nonwovens allow rapid access of the reaction components to the catalysts and of the products back into the reaction mixture.

7.1.5
Nanofiber Reinforcement

Fiber reinforcement is controlled primarily by the mechanical properties of the fibers, their axial ratio, and the mechanical coupling between the fibers and the matrix (e.g., see [126]). Electrospun nanofibers tend to display enhanced mechanical properties due to the self-organization controlling fiber formation during electrospinning. Furthermore, the axial ratio of the fibers, nanofibers with diameters of about 10 or 100 nm, can be 100 to 1000 times shorter than for fibers with diameters in the range of 10 to 100 μm. Due to their small diameters nanofibers cause only negligible refraction of light so that transparent matrices, which are reinforced by nanofibers, stay transparent, even if the refraction indices of matrix and fibers do not match. Finally, the large specific surface between nanofiber and matrix enhances relaxation processes, which improves the impact strength of the strengthened matrix.

7.2
Medicinal Applications

7.2.1
Tissue Engineering

Tissue engineering aims at artificially growing diverse types of tissues used to replace corresponding tissues destroyed by an accident or illness in the human body (e.g., see [127]). Cartilages, bones, skin, blood vessels, lymphatic vessels, lungs, and heart tissue are targets for this kind of reconstruction. One – in vitro or in vivo – approach in the field of tissue engineering is based on the use of scaffolds onto which cells or human body cells, respectively, can be settled. The scaffold has to facilitate anchorage, migration, and proliferation of the cells, to provide the three dimensional structure model of the tissue, and it has to fulfill a diverse range of requirements concerning biocompatibility, biodegradability, architecture, sterilizability, porosity, incorporation and release of drugs, mechanical properties, etc. Scaffolds based on nanofiber nonwovens and composed of synthetic biocompatible polymers or of natural ones, such as collagen, are highly promising. A major reason for this promise is that this approach allows the scaffolds to mimic the architecture of the extracellular matrix enclosing the cells in tissues. Furthermore, functional components such as growth factors can be incorporated via electrospinning.

7.2.2
Wound Healing with Nanofibers

An interesting application of nanofibers and especially electrospinning as a method to produce such nanofibers is in the treatment of large wounds, such as burns and abrasions. Observations are that these kinds of wounds heal particularly fast and without complications if they are covered by a thin web of nanofibers especially from biodegradable polymers (e.g., see [128]). Such nanowebs provide enough pores for an exchange of liquids and gases with the environment, but the pores are dimensioned in such a way that no bacteria can enter. Electrospun nanofiber nonwovens generally show very good adhesion even to moist wounds. Furthermore, the previously mentioned large surface of up to 100 m^2 per gram is highly beneficial for liquid adsorption and local release of drugs on the skin. In contrast to conventional wound treatment nanofibers also prevent scarring.

7.2.3 Transport and Release of Drugs/Drug Delivery

Nanofiber systems for the release of drugs or other functional compounds are generally not only of interest for wound healing or tissue engineering. Nanostructured systems for tumor therapy and also for other types of therapies like inhalation therapy or pain therapy are currently investigated worldwide (e.g., see [129,130]). In an ideal case, they have to fulfill versatile jobs in this function. The nanoobjects are supposed to protect the drugs in the case of systemic application from decomposition, e.g., in the blood circuit. Furthermore they should allow controlled release of the drug at a release rate as constant as possible over a longer period of time, adjusted depending on the field of application. They have to be able to permeate certain membranes or barriers, e.g., the blood–brain barrier, and they are supposed to concentrate the drug release only on the targeted body area.

Electrospun nanofiber may serve in this context as carriers for drugs and as controlled release agent. The domain of nanofibers loaded with drugs will most likely be of little importance for systemic therapy but of great importance for loco-regional therapy, i.e., the fibers are localized at the exact part of the body that is supposed to be treated with the carried drug. A currently developing field of application is inhalation therapy based on nanorods. The reason for this application is that the aerodynamic radius of such rods can be adjusted via the dimension of the rods in such a way that the drug carriers can be deposited at specific positions in the lung. It is well known from extensive experimental and theoretical studies correlated with the asbestos problem how fiber-shaped particles are deposited in the lung as a function of their axis ratio, length, radius, density, and surface structure [130, 131]. This knowledge can be used to place rod-shaped drug carriers at specific positions in the lung for loco-regional release. An advantage of using rodlike instead of spherical drug carriers is that the percentage of rods that remains in the lung after inhalation and is not exhaled is significantly higher than in the case of spherical particles. The treatment of tumors, metastases, pulmonary hypertension, and asthma, as well as the administration of insulin or other drugs via the lung are current research goals.

To be able to selectively adjust the aerodynamic diameters, the nanofibers fabricated by electrospinning have to be shortened to a defined axis ratio. This task can be achieved by laser or mechanical cutting. To control the aerodynamic radius via the density, highly porous fibers may be used. Inhalation therapy will have to be based on polymers which are biocompatible with particular emphasis on the specific reactions within the lung.

Acknowledgements The authors would like to thank Dr. Roland Dersch, Dr. Max von Bistram and Dr. Martin Graeser for their valuable support in the preparation of the manuscript. Financial support from the Deutsche Forschungsgemeinschaft (DFG) and the Volkswagen foundation is gratefully acknowledged.

References

1. Langer RM (1969) Integ Comp Biol 9:81
2. Wallenberger FT, Weston NE (2004) Natural fibers, plastics and composites. Springer, Heidelberg
3. Yang HH (1989) Aromatic high strength fibers. Wiley, New York
4. Cabasso I, Klein E, Smith JK (1976) Appl Polym Sci 20:2377
5. Folkes MJ (1982) Short fiber reinforced thermoplastics. Wiley, New York
6. Shin C, Chase GG (2004) AIChE J 50:343
7. Lee Y, Wadsworth LC (1992) Polymer 33:1200
8. Hagewood J, Wilkie A (2003) Nonwovens World 12:69–73
9. Huang ZM, Zhang YZ, Kotaki M, Ramakrishna S (2003) Compos Sci Technol 63:2223
10. Dersch R, Greiner A, Wendorff JH (2004) In: Schwartz JA, Contesen CJ, Putger K (eds) Dekker Encyclopedia of Nanoscience and Nanotechnology. Marcel Dekker, New York, p 2931
11. Li D, Xia Y (2004) Adv Mater 16:1151
12. Greiner A, Wendorff JH (2007) Angew Chem Int Edit 119:5750
13. Reneker DH, Yarin AL, Zussman E, Xu H (2007) Adv Appl Mechanics 41:44
14. Barrero A, Loscertales IG (2007) Ann Rev Fluid Mech 39:89
15. Marin AG, Loscertales IG, Marquez M, Barrero A (2007) Phys Rev Lett 98:014502
16. Lehn M (1990) Angew Chem Int Edit 11:1304
17. Vriezema DM, Aragones AC, Elemans AAW, Cornelissen JLM, Rowan AE, Nolte RJM (2005) Chem Rev 105:1445
18. Earnshaw S (1842) Trans Camb Phil Soc 7:97
19. Reneker DH, Yarin AL, Fong H, Koombhongse S (2000) J Appl Phys 87:4531
20. Yarin AL, Koombhongse S, Reneker DH (2001) J Appl Phys 90:4836
21. Reznik SN, Yarin AL, Theron A, Zusmann E (2004) J Fluid Mech 516:349
22. Yarin AL, Zussman E (2004) Polymer 45:2977
23. Larrondo L, Manly RSJ (1981) J Polym Sci Polym Phys 19:921
24. Xu H, Yarin AL, Reneker DH (2003) Polym Prepr 44:51
25. de Gennes P (2003) J Chem Phys 60:5030
26. Warner SB, Buer A, Ugbolue SC, Rutledge GC, Shin MY (1998) Nat Textile Center Annual Rep No 83
27. Lord Rayleigh X (1882) Philos Mag 44:18
28. Zeleny J (1917) Phys Rev 10:1
29. Taylor G (1964) Proc R Soc London A 280:383
30. Yarin AL, Koombhongse S, Reneker DH (2001) J Appl Phys 89:3018
31. Hohman MM, Shin M, Rutledge G, Brenner MP (2001) Phys Fluids 13:2201
32. Hohman MM, Shin M, Rutledge G, Brenner MP (2001) Phys Fluids 13:2221
33. Shin YM, Hohmann MM, Brenner M, Rutledge GC (2001) Polymer 42:9955
34. Fridrikh SV, Yu JH, Brenner MP, Rutledge GC (2001) Appl Phys Lett 78:1149
35. Fridrikh SV, Yu JH, Brenner MP, Rutledge GC (2001) Phys Rev Lett 90:144502
36. Yarin AL, Kataphinan W, Reneker DH (2005) J Appl Phys 98:064501

37. Stelter M, Brenn G, Yarin AL, Singh RP, Durst F (2000) J Rheol 44:595
38. Holzmeister A, Greiner J, Wendorff JH (2007) Eur Polym J 43:4859
39. He JH, Wan YQ, Yu JY (2004) J Nonlinear Sci Numer Simul 5:253
40. Wan YQ, He JH, Yu JY, Wu Y (2006) J Appl Polym Sci 103:3840
41. Fridrikh SV, Yu JH, Brenner MP, Rutledge GC (2003) Phys Rev Lett 90:144502
42. Baumgarten K (1971) J Colloid Interf Sci 36:71
43. Theron SA, Zussman E, Yarin AL (2004) Polymer 45:2017
44. Hou H, Jun Z, Reuning A, Schaper A, Wendorff JH, Greiner A (2002) Macromolecules 35:2429
45. Koombhongse S, Liu WX, Reneker DH (2001) J Polym Sci B 39:2599
46. Ma M, Gupta M, Li Z, Zhai L, Gleason KK, Cohen R, Rubner MF, Rutledge GC (2007) Adv Mater 19:255
47. Lord Rayleigh X (1892) Phil Mag 34:145
48. Tomotika S (1935) Proc R Soc London A 150:322
49. Higgins JS, Tambasco M, Lipson JEG (2005) Prog Polym Sci 30:832
50. McCann J, Marquez M, Xia Y (2006) J Am Chem Soc 128:1436
51. Casper CL, Stephens JS, Tasi NG, Rabolt JF (2004) Macromolecules 37:573
52. Megelski S, Stephens JS, Bruce CD, Rabolt JF (2002) Macromolecules 35:8456
53. Bognitzki M, Czado W, Frese T, Schaper A, Hellwig M, Steinhart M, Greiner A, Wendorff JH (2001) Adv Mater 13:70
54. Vogt M (2006) Diploma Thesis, Philipps University, Marburg
55. Weisbuch C, Vinter B (2007) Quantum Semiconductor Structures. Academic Press, New York
56. Struik LCE (1977) Polym Eng Sci 17:165
57. Efremov MY, Olson EA, Zhang M, Zhang Z, Allen LH (2004) Macromolecules 37:4607
58. Efremov MY, Olson EA, Zhang M, Zhang Z, Allen LH (2004) Phys Rev Lett 91:085703
59. Agarwal S, Puchner M, Greiner A, Wendorff JH (2005) Polym Int 54:1422
60. Kim JS, Reneker DH (1999) Polym Eng Sci 39:849
61. Strobl G (1997) The Physics of Polymers. Springer, Berlin
62. Kim JS, Lee DS (2000) Polym J 32:616
63. Dersch R, Liu T, Schaper A, Greiner A, Wendorff JH (2003) Polym Sci A 41:545
64. Stephens JS, Chase DB, Rabolt JF (2002) Macromol 37:877
65. Dersch R (2002) Diploma Thesis, Philipps University, Marburg
66. Chen Z, Foster MD, Zhou W, Fong H, Reneker DH (2001) Macromolecules 34:6156
67. Jaeger R, Schönherr H, Vansco GJ (1996) Macromolecules 29:7634
68. Krause S, Dersch R, Wendorff JH, Finkelmann H (2007) Macromol Rapid Commun 28:2062
69. Ward IM, Sweeney J (2004) An introduction to the mechanical properties of solid polymers. Wiley, Weinheim
70. Thumser DA, Bergmann R (2005) Material Werkst 36:204
71. Gu S-Y, Wu Q-L, Ren J, Vansco GJ (2005) Macromol Rapid Commun 26:716
72. Bellan M l, Kameoka J, Craighhead HG (2005) Nanotechnology 16:1095
73. Lee SH, Tekman C, Sigmund WM (2005) Mater Sci Eng A 398:77
74. Zusssman E, Theron A, Yarin AL (2003) Appl Phys Lett 82:973
75. Li D, Wang Y, Xia Y (2003) Nano Lett 3:1167
76. Kameoka J, Craighead HG (2003) Appl Phys Lett 83:371
77. Sun D, Chang C, Li S, Lin L (2006) Nano Lett 6:839
78. Belardi J (2007) Diploma Thesis Marburg
79. Kidoaki S, Kwon IK, Matsuda T (2004) Biomaterials 26:37

80. Fagerlund G (1973) Determination of specific surface by the BET method. Springer, Dordrecht
81. Tomadakis MM, Sitirchos SV (1991) AIChE J 37:74
82. Tomadakis MM, Sitirchos SV (1993) AIChE J 39:397
83. Tomadakis MM, Sitirchos SV (1993) J Chem Phys 98:616
84. Tomadakis MM, Sitirchos SV (1991) AIChE J 37:1175
85. Gorham WF (1964) J Polym Sci A 1(4):3027
86. Hellen EKO (1997) J Appl Phys 81:6425
87. Maze B, Tafreshi VH, Wang Q, Pourdeyhimi B (2007) J Aerosol Sci 38:550
88. Matthews JA, Wnek G, Simpson DG, Bowlin GL (2002) Biomacromolecules 3:232
89. Placke D (2007) PhD Thesis, Philipps University, Marburg
90. Sun Z, Zussman A, Yarin AL, Wendorff JH, Greiner A (2003) Adv Mater 15:1929–1932
91. Reznik SN, Yarin AL, Zussman E, Bercovici L (2006) Phys Fluids 18:062101/01-062101/13
92. Yarin AL, Zussman E, Wendorff JH, Greiner A (2007) J Mater Chem 17:1–16
93. Greiner A, Wendorff JH, Yarin AL, Zussman E (2006) Appl Microbiol Biotechnol 71:387
94. Yu JH, Fridrikh SV, Rutledge GC (2004) Adv Mater 16:1562–156
95. Lopez-Herrera JM, Barrero A, Lopez A, Loscertales IG, Marquez M (2003) J Aerosol Sci 34:535–552
96. Hou H, Jun Z, Reuning A, Schaper A, Wendorff JH, Greiner A (2002) Macromolecules 35:2429–2431
97. Bognitzki M, Hou H, Ishaque M, Frese T, Hellwig M, Schwarte C, Schaper A, Wendorff JH, Greiner A (2000) Adv Mater 12:637–640
98. Caruso R, Schattka JH, Greiner A (2001) Adv Mater 13:1577–1579
99. Sun Z, Zeng J, Hou H, Wickel H, Wendorff JH, Greiner A (2005) Prog Colloid Polym Sci 130:15–19
100. Flory PJ (1969) Statistical Mechanics of Chain Molecules. Butterworth-Heinemann, London
101. de Gennes P (1980) Scaling Concepts in Polymer Physics. Cornell University Press, Cornell, NY
102. Larrondo L, Manley RS (1981) J Polym Sci Polym Phys 19:909
103. Kim JS, Lee D-S (2001) Polymer J 32:616
104. Lee S, Obendorf SK (2006) J Appl Polym Sci 102:3430
105. Dalton PD, Klinkhammer K, Salber J, Klee D, Möller M (2006) Biomacromolecules 7:686
106. M Becker (2006) PhD Thesis, Philipps University, Marburg
107. Stoiljkovic A, Ishaque M, Justus U, Hamel L, Klimov E, Heckmann W, Eckhardt B, Wendorff JH, Greiner A (2007) Polymer 48:3974
108. Matthews JA, Wnek GE, Simpson DG, Bowlin GI (2002) Biomacromolecules 3:232
109. Ohkawa K, Cha D, Kim H, Nishida A, Yamamoto H (2004) Macromol Rapid Commun 25:1600
110. Stephens JS, Fahnestock SR, Farmer RS, Kiick KL, Chase DB, Rabolt JF (2005) Biomacromolecules 6:1405
111. Zheng M, Skomski R, Liu Y, Sellmyer DJ (2000) J Phys Condens Mater 12:L497
112. Martin JI, Costa-Kramer JL, Briones F, Vicent JL (2000) J Magn Magn Mater 221:215
113. Sorop TG, Nielsch K, Göring P, Kröll M, Blau W, Wehrspohn RB, Gösele U, de Jongh LJ (2004) J Magn Magn Mater 272–276:1656
114. Graeser M, Greiner A, Wendorff JH (2007) Macromolecules 40:6032

115. Graeser M (2007) PhD Thesis, Philipps University, Marburg
116. Tan ST, Wendorff JH, Pietzonka C, Jia ZH, Wang GQ (2005) ChemPhysChem 6:1461
117. Wang M, Singh H, Hatton TA, Rutledge GC (2004) Polymer 45:5505
118. Bognitzki M, Becker M, Graeer M, Massa W, Wendorff JH, Schaper A, Weber D, Beyer A, Gölzhäuser A, Greiner A (2006) Adv Mater 18:2384
119. Graeser M, Greiner A, Pietzonka C, Massa W, Wendorff JH (2007) Adv Mater 19:4244
120. Schreuder-Gibson HL, Gibson P, Senecal K, Sennett M, Walker J, Yeomans W, Ziegler D, Tsai PP (2002) J Adv Mater 34:44
121. Gibson P, Schreuder-Gibson H, Rivin D (2001) Colloid Surf A 187/188:469
122. Hajara MG, Metha K, Chase GG (2003) Sep Filtr Technol 30:74
123. Maze B, Vahedi Tafreshi H, Wang Q, Pourdwyhimi B (2007) J Aerosol Sci 38:550
124. Stasiak M, Röben C, Rosenberger N, Schleth F, Studer A, Greiner A, Wendorff JH (2007) Polymer 48:5208
125. Berghoef MM, Vansco GJ (1999) Adv Mater 11:1362
126. Boudriot U, Dersch R, Greiner A, Wendorff JH (2006) Artif Org 30:785
127. Boland ED, Wnek GE, Simpson DG, Palowski KJ, Bowlin GL (2001) J Macromol Sci Pure Appl Chem 38:1231
128. Sanders EH, Kloefkorn R, Bowlin GL, Simpson DG, Wnek GE (2003) Macromolecules 36:3803
129. Zeng J, Aigner A, Czubayko F, Kissel T, Wendorff JH, Greiner A (2005) Biomacromolecules 6:1484
130. Timbrell V (1965) Ann NY Acad Sci 132:255
131. Harris RL, Fraser DA (1976) Am Industr Hygien Assoc J 37:73

Subject Index

Adenosine 5′-monophosphate (AMP) 7
Alkylthiols 9
Amino acids, isoindole derivatives 9
Amorphous polymers 138
Amphiphiles, bilayer-forming 4
Amphiphilic rods 96
Amyloid-like fibrils 29
Artificial peptides 27, 30
ATP, cyanine dye 10
Au nanoparticles, lipophilic, diaminopyridin (DAP) 55
Axisymmetric instability 125
Azobenzene 44

Band gap engineering, supramolecular 15
Beaded fibers 124, 134
Bending instability 125
Bioconjugate amphiphilic rods 96
Biomedical engineering 52
Biomimetic amphiphilic rods 96
Biopolymers, spinning 158
Bovin serumalbumin 159
Buckling 132

Carbohydrate-conjugate molecular wedge 101
Casein 159
Catalysis 165
Cellulose 159
Cellulose acetate (CA), electrospun, organic solvents 156
–, electrospinning 119
Charge transfer 1
Chitosan 159
Coalescer filters 110
Cobalt nanowires 163
Co-electrospinning 108, 151
Coiled-coil motif/structure 27, 56
Collagen 62, 158, 165
Collagen-like triple helix foldamer 62
Concanavalin A 100
Conjugate rods, lateral chains 89
Controlled release agent 167
Copper nanofibers 163
Copper-phtalocyanine tetrasulfonate 11
Core shell fibers 152
Cylindrical micelles 69

Dendron–rod–coil molecules 69, 80
Dialkyldimethylammonium 7
Dihexadecyl sulfosuccinate 14
DNA templated self-assembly 98
Dodeca-*para*-phenylene rod 82
Domain-swapped open aggregate 60
Droplet deformation 120
Drug carriers 167
Drugs/drug delivery, transport/release 167
Dumbbell-shaped molecules 69, 82

Earnshaw theorem of electrostatics 113
Electrospinning 108, 111, 114
–, organic solvents 156
–, reactive 157
–, reduced electrode distances 149
–, water soluble polymers 157
Electrospun nanofibers 164
2-Ethylhexyl methacrylate (EHMA) 81

Fiber cross sections 132
Fiber diameter, reduction 110
Fiber formation/architectures 108, 115
Fiber reinforcement 110
Filters 165
–, coalescer 110
Fluid filtration 110
Fractal surface structures 111
Fragmentation, linear segments 134

Gas filtration 110
Guanosine 5′-monophosphate (GMP) 7

Helicates 12
Helix structure 27
Helix-based self-assembly 56
Hexa-*para*-phenylene (HP) rod 74
High precision deposition electrospinning (HPDE) 150
Hollow fibers, via templates 152
Honeycomb morphology 17
Hydrogen bonding 1
Hydroxypropylcellulose (HPC), electrospun 157

Inhalation therapy 167
Iron oxide, superparamagnetic 161
Isoindole 9

Jet, looping/straight 117
Jet branching 127
Jet initiation, droplet deformation 120

Light-harvesting hydrogels 8
Lipase 159
Lipophilic bridging ligands, metal ions 18
Luciferase 159

Macrocycles 69, 77
Man-made fibers 109
Medicinal applications 166
2-Mercaptoethanesulfonic acid (MES) 9
Mesoscopic supramolecular assembly 1
Metal complex 1
Metal nanofibers 160, 162
Metal oxide nanofibers 160, 162
Metalloamphipihles 5
Mixed valence complexes, halogen-bridged 12
Molecular pairing 7
Molecular recognition 5
Molecular rods 69
–, one-dimensional assemblies in solutions 72
Molecular wires 12
Myelinated axon 2

Nanobiotechnology 27
Nanofibers 1, 69, 108
–, biological molecular components 9
–, diameters 128
–, molecular pairing 7
–, morphology, supramolecular control 20
–, organic/inorganic, self-assembly in organic media 12
–, partial crystalline 139
–, porous 136
–, properties 128
–, reinforcement 165
–, self-assembly, classical amphiphiles 3
–, –, β-sheet-based 29
–, supramolecular 1
–, topologies 134
Nano-templates 53
Nanowires 53
–, barbed 111
–, solvatochromic 15
Neurons 2
NIPAM 51
Nonwovens 108
–, electrospun nanofibers 142
–, fiber arrangement 143
–, heterogeneous 144
–, mechanical properties 148
Nucleic acids 9

Oligo(*para*-phenylene) rod 74
Oligo(*para*-phenyleneethynylene) 93
Oligo(*para*-phenylenevinylene)s 90
Oligopeptides 9
One-dimensional nanostructures 69
Ortho-phthalaldehyde (OPA) 9

Pain therapy 167
Palladium 160
PEG-*block*-oligopeptide 46
Peptide nanofibers 27
–/synthetic polymers, hybridization 46
Peptidomimetics 30
Perylenediimide 11
Photo-controlled peptide self-assembly 41
Platinum 160
PNIPAM 51, 55
Poly(acrylic acid) (PAA), electrospun 157
Poly(acrylic amide), electrospun 157
Poly(ε-caprolactone) (PCL) 159
Poly(dicyclopentadien) fibers 157
Poly(ethylene oxide) (PEO) coils 74
Poly(ethylene-*co*-vinylacetate) (PEV), electrospun, organic solvents 156

Poly(ethylenterephthalate) (PET), electrospun, organic solvents 156
Poly(hexamethylenterephthalate-*co*-hexamethylen-2,6-naphthalate) (P(HT-co-HN)), electrospun, organic solvents 156
Poly(hydroxybutyrate)s 159
Poly(*p*-phenylenterephthalate) (PPTA), electrospun, organic solvents 156
Poly(propylene oxide) (PPO) coils 75
Poly(trimethylenterephthalate) (PTT), electrospun, organic solvents 156
Poly(vinylpyrrolidon) (PVP), electrospun 157, 158
Polyacrylonitrile (PAN), electrospun, organic solvents 156
Polyamide 12 155
Polyamides (PA), electrospun, organic solvents 156
Polybenzimidazol (PBI), electrospun, organic solvents 156
Polycaprolactone polyamide 12 155
Polycarbonate (PC), electrospun, organic solvents 156
Polyelectrolytes, electrospun 157
Polyethylene oxide (PEO), electrospun 157
Polyethylene terephthalate 155
Polyimides (PI), electrospun, organic solvents 156
Polylactide nanofibers 159
–, annealing 119
Polymer hybrids 160
Polymethylmethacrylate (PMMA), electrospun, organic solvents 156
Polyparaxylylene (PPX) 153
Polypropylene 155
Polystyrene (PS), electrospun, organic solvents 156
Polystyrene latex particles 158
Polyurethanes (PU) 155
–, electrospun, organic solvents 156
Polyvinylacetate (PVA), electrospun 157
Polyvinylbutyral (PVB), electrospun, organic solvents 156
Polyvinylchloride (PVC), electrospun, organic solvents 156
Polyvinylidenfluoride (PVDF), electrospun, organic solvents 156
Pore structures 146
Porosity 146
Porous fibers 134
Precision deposition 150
Precision electrospinning 108
Proteins 9

Reactive electrospinning 157
Reinforcement 110, 165
Release agent 167
Rhodium 160
Rod–coil molecules 69, 72
–, macrocycles 78
Rods, amphiphilic 96

Self-assembly 1, 69
–, 1D conjugated electronic structures 14
–, fiber peptides 58
–, interfaces 17
–, supramolecular nanofibers, aqueous media 5
Sergeant-and-soldiers coassembly 93
β-Sheet structure 27
Silk 108, 159
Solvatochromic nanowires 15
Spider webs 108
Spiked fibers 124
Spin conversion directed by self-assembly 22
Spin crossover 1, 20
Spindle-type disturbances 11
Spinning, biopolymers 158
–, polymer melts 154
–, technical polymers 154
Stereo-honeycomb nanoarchitectures 17
Stimuli-responsive self-assembly 41
Streptavidin 55
Stress-strain 142, 148
Sugar-based rod–coil 76
Supramolecular objects 69

Tape-like nanofiber 45
Templates 164
Thiacarbocyanine dye 9
Tissue engineering 166
Triazoles, bridging ligands 18
Tubes by fiber templates (TUFT) 152, 164
Tumor therapy 167

Vancomycin 7

Wedge-type molecules 87
Wound healing 158, 166

The manufacturer's authorised representative in the EU is Springer Nature Customer Service Centre GmbH, Europaplatz 3, 69115 Heidelberg, Germany. If you have any concerns regarding our products, please contact ProductSafety@springernature.com

Printed and bound by CPI Group (UK) Ltd, Croydon, CR0 4YY
02/10/2026
02226631-0003